Design and Analysis of Reliability Studies

Design and Analysis of Reliability Studies

The statistical evaluation of measurement errors

GRAHAM DUNN
Institute of Psychiatry, London

A CHARLES GRIFFIN TITLE

OXFORD UNIVERSITY PRESS
New York

Edward Arnold
A division of Hodder & Stoughton
LONDON MELBOURNE AUCKLAND

© 1989 Graham Dunn

First published in Great Britain 1989

British Library Cataloguing in Publication Data

Dunn, Graham
   Design and analysis of reliability studies.
   1. Science. Measurement. Errors. Statistical analysis
   I. Title
   530.8'01'5195

   ISBN 0-85264-297-0

Typeset in 10/11 Times by Blackpool Typesetting Services Ltd., Blackpool, UK.
Printed and bound in Great Britain for Edward Arnold, the educational, academic and medical publishing division of Hodder and Stoughton Limited, 41 Bedford Square, London WC1B 3DQ by Richard Clay plc, Bungay, Suffolk.

Published in the USA by
Oxford University Press
200 Madison Avenue, New York, NY 10016

Library of Congress Cataloguing-in-Publication Data

Dunn, G. (Graham), 1949–
   Design and analysis of reliability studies: the statistical
   evaluation of measurement errors/Graham Dunn.
      p.   cm.
   Includes bibliographical references.
   ISBN 0-19-520704-1
   1. Error analysis (Mathematics)   I. Title.
QA275.D86   1989
511'.43—dc20   89-37696
                                                         CIP

Printed in Great Britain

Contents

Preface

This book is a general and practically orientated introduction to the dependability or reliability of measurements, the latter ranging from nominal classifications through simple ordinal rating scales to data assumed to be measured on an interval scale. Although primarily intended to be used by behavioural, medical and social scientists, it should be useful to physical scientists and biologists. One of the implicit themes of the book is that the measurement models and the statistical methods used in the assessment of precision, bias and reliability are common to virtually all fields of scientific endeavour but, unfortunately, these methods are often developed by one discipline in isolation or are independently developed by several disciplines, each being ignorant of the progress of the others. The other group of potential readers of this text are, of course, applied statisticians who may be concerned with giving advice about the measurement of reliability but who will not be specialists in the field and therefore will be unaware of, or not have access to, the wide-ranging literature in this field.

There are essentially three independent parts to this text. The first comprises Chapters 1–3, the second is Chapter 4, and the last comprises Chapters 5–7. These seven chapters are then supplemented by four technical appendices and a bibliography. The first three chapters cover the concepts required for the study of the rest of the text. They are measurement models (Chapter 1), coefficients of agreement and disagreement (Chapter 2), and indices of reliability (Chapter 3). Chapter 4 more or less stands alone and is concerned with the design of a study to assess the performance of a measuring device or group of measuring instruments. Finally, the last part of the book covers inferential methods that are useful in the analysis of data generated in a reliability study. Chapters 5 and 6 cover techniques appropriate for interval data, whilst Chapter 7 concentrates on those more appropriate for categorical, including ordinal, measurements.

The aim in writing this text has been to produce a guide to a wide variety of ideas and methods that can be of use in a reliability study, including enough theoretical material to enable the user of the relevant techniques to understand why and how they can be used, but not concentrating on the formal mathematical proofs and derivations which might be found in a more advanced statistical monograph. The level of statistical competence expected of the reader is that covered by most elementary statistics courses, including analysis of variance, linear regression, and chi-square tests for two-way contingency tables. It is vital for the use of most of the present book, however, that the reader is reasonably familiar with the concept of and use of expected values. This area of statistics is not usually covered in elementary courses or text books and, therefore, is here explained in considerable detail in Appendices 1–3. It is recommended that a reader who is unfamiliar with expected values should begin the study of this text by reading these three appendices.

It is taken for granted by the author that a user of this text will wish to analyse his or her own data through the use of a computer, but will not necessarily have the skill or time to write purpose-built programs. Appendix 4 provides a brief guide to commercially available software that might be useful in tackling the statistical problems described in the text. It is not, however, intended to be a complete guide. It is simply a description of some of the readily available software which the author has found to be useful in the preparation of this book.

Graham Dunn
July 1988

Acknowledgements

Many of the examples of analyses described in this text are dependent on data provided by the author's colleagues at the Institute of Psychiatry. He is particularly grateful to Dr Stuart Turner (CAT scan data), Dr Michael Craggs (EEG measurements), Professor David Hand (referee's assessments of conference abstracts), and to Dr Glyn Lewis (psychiatric interview and questionnaire responses). The author accepts full responsibility for any errors or misconceptions concerning these data sets but, does, of course, assume that there are none! The author and publisher would also like to thank authors and publishers for permission to reproduce other material that is used in the text. Individual items are acknowledged at the appropriate places in the book.

The author would also like to thank the editor of this series, Professor Brian Everitt, for suggesting the writing of this book and for commenting on preliminary drafts of each of the chapters. Finally, he would like to express his appreciation to Mrs Bertha Lakey for patiently typing the manuscript.

1

An Introduction to Measurement Models

1.1 Introduction

Imagine a rather obsessional cook who wishes to test the 'accuracy' of his kitchen scales. The basic features of these scales incorporate a large pan on the top and a pointer and scale beneath to indicate the weight of the contents of the pan. The only other feature of interest is a lever which enables him to re-set the pointer's zero position when the pan contains nothing. The scale indicates weights from zero to 3 kg in 20 g intervals. It is unlikely, therefore that the cook could estimate the value of a single reading any closer than to the nearest 10 g. The cook takes a packet of dried fruit from his kitchen cupboard and notes that the packers have claimed that it weighs 500 g. This will be assumed, for the time being, to be the packet's true weight. He takes ten repeated measurements of the fruit's weight on his scales and obtains the following readings (in g):

$$490, 520, 500, 520, 470, 490, 490, 490, 500, 480.$$

The arithmetic mean of these recordings is 495 and their variance is 250 (standard deviation 15.8). The arithmetic mean gives an indication of the weight that would be obtained on average if the weighing process could be repeated indefinitely, and the difference between this value and the packet's 'true' weight of 500 g indicates, on the basis of the results of this simple experiment at least, that the scales are likely to underestimate the weight of an item of food. The scales have an estimated *bias* of −5 g. The variance (or equivalently the standard deviation) of the readings gives the cook an idea of the *precision* of his kitchen scales. A high variance is regarded as an indicator of low precision. The standard deviation of the measurements is often referred to as the *standard error of measurement* of the scales.

Now, having thought about the results of the first experiment, the cook realises that the bias observed may have been caused by the way in which the zero had been set prior to the experiment. It could quite easily have been due to chance, of course, but in any case he decides to carry out a further experiment. In this experiment he weighs the same packet of dried fruit ten times, but prior to each weighing he re-sets the machine's pointer to zero. The ten readings now obtained are the following:

480, 500, 510, 450, 480, 500, 490, 460, 480, 450.

Their arithmetic mean is 480 with variance 444.4 (standard error of measurement 21.1).

These two simple experiments have been used to illustrate the two basic concepts of bias and precision. They also illustrate that both bias and precision are dependent on the way a measuring instrument is used. A particular source of error (such as re-setting the zero on the kitchen scales) can contribute to bias or to lack of precision. In the cook's first experiment the zero was set only once and therefore the influence of this setting was common to all readings (that is it was a potential source of bias). In the second experiment the same re-setting operation was carried out before each reading was taken, and in this case the procedure resulted in lower precision. A systematic error in the first experiment has become a random or haphazard one in the second. The change in bias is not explained by the change in experimental design, however, and further experiments would be needed to investigate the source of this bias.

Having learnt a little about errors of measurement, how should the cook now proceed to use his kitchen scales? Higher precision can be gained by just resetting the scales' zero once. This may or may not be a good procedure depending on how easily or how quickly the zero drifts away from its pre-set position during use. This does not solve the problem of bias, however. This might be solved by first weighing an empty container and then adding the required amount of food. In this case the bias ought to be common to both readings and would disappear on subtraction to obtain the weight of the food alone. The improvement might be an illusion, however, the variance of the errors of the difference will be double that of the errors of an individual reading (see Appendix 2). However, this problem can be solved quite easily by simply repeating the measurement process several times and then calculating the mean of the differences.

Before leaving the kitchen for the scientific laboratory it might be useful to consider the implications of the errors of measurement introduced through the use of the above kitchen scales. The conclusion ought to be that the likely size of the errors introduced by these scales when weighing foods such as sugar, apples or potatoes is trivial. A 25 g error should be of no consequence to the cook, however obsessional he is. This can be illustrated by introducing the *coefficient of variation* as a measure of relative precision. The sample coefficient of variation is simply obtained by dividing the standard error of measurement by the corresponding arithmetic mean. The coefficients of variation for the two experiments above are 0.03 and 0.04, respectively. (It can be multiplied by 100 to be expressed as a percentage.) If, however, the mean weight given in the first experiment were 5 g instead of about 500 g, but with the same standard error of measurement, then the coefficient of variation would be 3.16 (or 31.6%). If the cook were to weigh very small amounts of spices, for example, then the above standard errors of measurement are no longer trivial. The difference between 1 g and 30 g of hot chilli powder is enormous! Clearly one needs scales of much higher precision for the measurement of commodities such as spices. In other words, the accuracy required of any measurement instrument is dependent on the context of its use; a standard error (or bias) of a particular size may be trivial in one context but absolutely vital in another.

1.2 Statistical models for physical measurements

When one moves from the kitchen to a physics or chemistry laboratory then one wishes to go further than the rough and ready analysis presented above. Ideally one should be able to understand the measurement process well enough to be able to make inferences about the likely errors in any given measurement or estimate of a given characteristic. What is its standard error, for example? How confident can one be that the characteristic's value falls within a given range? How many repeated measurements need to be made to achieve an estimate with a given precision? How many measurements are needed to convince a sceptic that a characteristic has a value that differs from that predicted by a given theory?

It is a truism in physics that a measurement or estimate is worthless unless it is accompanied by an estimate of its precision (in the form of a range of possible values, a confidence interval or simply its standard error, for example). To illustrate this point, Taylor (1982) describes a problem similar to one that is said to have been solved by Archimedes. One has to decide whether a crown is made of 18-karat gold (with a known density of 15.5 g/cm^3) or a cheap alloy (with a known density of 13.8 g/cm^3). Two independent estimates of the crown's density are 13.9 g/cm^3 and 15 g/cm^3. On their own one cannot tell which of the two estimates is the most reliable nor can one make the required decision. The data as they stand are worthless. If, however, more information is provided, then one can make use of these estimates. The expert who provided the first estimate also stated that it is highly probable that the true density falls between 13.7 g/cm^3 and 14.1 g/cm^3. The expert providing the second estimate stated that he thought that the true density must lie between 13.5 g/cm^3 and 16.5 g/cm^3. Providing that these two experts could justify their conclusions concerning the above intervals (and in practice they would need to) then one can justifiably conclude that the crown is made of alloy.

The simplest statistical model for a series of measurements of weights or densities has the following form:

$$X_{ij} = \tau_i + e_{ij} \qquad (1.1)$$

In this model, X_{ij} represents the jth measurement on the ith object ($i = 1, 2, \ldots, I$; $j = 1, 2, \ldots, J$), τ_i is the 'true' value for object i and e_{ij} is the measurement error associated with X_{ij}. Those readers unfamiliar with expected values are referred to Appendix 1. Note that

$$E_j(X_{ij}) = \tau_i \qquad (1.2)$$

$$\text{and } E_{i,j}(X_{ij}) = E_i(\tau_i) = \mu \qquad (1.3)$$

where μ is the mean for the whole population of possible measurements and the subscripts under the expectation operators refer to the population of values over which the expectations are being taken. Note that a 'true' value is defined in terms of expectations (Equation (1.2)), and it may not correspond to the truth in a Platonic sense. If the measurement instrument is biased, then, in this model, the bias will be confounded with the τ_is. The error terms (the e_{ij}) are usually assumed to be independent of one another (between and within objects) and to be independent of the expected values (the τ_i).

If there are two measuring instruments, then one can extend the model to the following. For test or instrument 1:

$$X_{ij1} = \tau_{i1} + e_{ij1} \qquad (1.4)$$

and for test or instrument 2:

$$X_{ij2} = \tau_{i2} + e_{ij2} \qquad (1.5)$$

where the subscripts 1 and 2 refer to the two measuring instruments in an obvious way, i refers to the ith object and j to the jth measurement made on the ith object ($i = 1, 2, ..., I; j = 1, 2, ..., J$). Now, if $\tau_{i1} = \tau_{i2}$ for all of the i objects being measured and also if the variances of the error terms e_{ij1} and e_{ij2} are the same (that is the standard errors of measurements for the two instruments are the same), then the instruments are referred to as being *parallel*. If there are K different measuring instruments then, for the kth instrument ($k = 1, 2, ..., K$):

$$X_{ijk} = \tau_{ik} + e_{ijk} \qquad (1.6)$$

Note that there has still not been any reference to the Platonic true values of the characteristic being measured by these instruments; the bias of an instrument is not absolute but always relative to an alternative measuring instrument.

Before moving on to more complex measurement models, the ideas already introduced will be clarified and expanded. Returning to the cook's first experiment, it is seen that only one measuring instrument is involved and also only one object is being weighed. The same applies to his second experiment. These two experiments lead to measurements that can be described by the following simple model:

$$X_j = \tau + e_j \qquad (1.7)$$

The subscripts i and k have been dropped because they are not necessary. The constant τ can be estimated by the arithmetic mean of the X_j and the error variance by the variance of the Xs. If the error terms are now assumed to be normally distributed with mean 0 and variance σ_e^2 (estimated by the above sample variance), then it is straightforward to produce, say, a 95% or 99% confidence interval for τ.

In the earlier discussion of the weights of the packet of dried fruit, it was mentioned that the weight stated on the packet was 500 g. Whether or not this is regarded as actually being the true weight of the packet, Equation (1.7) could be modified in the following way:

$$X_j = 500 + \tau + e_j \qquad (1.8)$$

In this equation τ is now a measure of bias relative to the fixed *reference* value of 500 g.

Table 1.1 shows the measurements obtained in a third, slightly more sophisticated experiment with the kitchen scales. Here a sample of twenty objects (packets containing potatoes) were weighed separately. The pointer on the scales was then re-set to zero and the 20 weights measured a second time. Here the appropriate statistical model might be

$$X_{ik} = \tau_{ik} + e_{ik} \qquad (1.9)$$

Table 1.1 Repeated measurements of the weights of several packets of potatoes (in grammes)

Packet (i)	Time 1 (X_{i1})	Time 2 (X_{i2})	Difference ($X_{i1} - X_{i2}$)
1	560	550	10
2	140	140	0
3	1680	1710	-30
4	1110	1090	20
5	1060	1040	20
6	280	250	30
7	620	600	20
8	830	800	30
9	690	690	0
10	1210	1170	40
11	2300	2260	40
12	1880	1850	30
13	2000	2000	0
14	2360	2300	60
15	1670	1720	-50
16	1230	1230	0
17	1370	1390	-20
18	1750	1710	40
19	1730	1710	20
20	1680	1640	40
		Mean	15
		Variance	721

Here, since there is only one measurement at each time, $j = 1$ for all observations, and it has therefore been ignored. Note also that in this case the subscript k ($k = 1, 2$) is referring to the time of the measurement; there is still only one measurement instrument involved. In this example there appears to be a problem. There are forty parameters given by Equation (1.9) ($\tau_{11}, \tau_{12}, \tau_{21}, \tau_{22}, \ldots, \tau_{20,1}, \tau_{20,2}$) and there are only forty observations. The errors are confounded with the true values. However, if it is assumed that

$$\tau_{i1} - \tau_{i2} = \alpha_{12} \qquad (1.10)$$

for all i, then Equation (1.9) can be replaced by

$$X_{i1} - X_{i2} = \alpha_{12} + (e_{i1} - e_{i2}) \qquad (1.11)$$

The constant relative bias α_{12} can be estimated from the arithmetic mean of the twenty difference scores, $X_{i1} - X_{i2}$, and the variance of the error terms by dividing the variance of the difference scores by two. This follows from Equation (1.11):

$$\mathrm{Var}(X_{i1} - X_{i2}) = \mathrm{Var}(e_{i1} - e_{i2})$$

$$= \mathrm{Var}(e_{i1}) + \mathrm{Var}(e_{i2})$$

$$= 2\sigma_e^2 \qquad (1.12)$$

As before it has been assumed that the error terms are independent of each other and of the parameters of the statistical model, and also that the error variance (precision) is the same at the two occasions.

Table 1.2   Subjective estimates of the lengths (to nearest 1/10 inch) of 15 pieces of string

String	'True' length	Graham	Brian	David
A	6.3	5.0	4.8	6.0
B	4.1	3.2	3.1	3.5
C	5.1	3.6	3.8	4.5
D	5.0	4.5	4.1	4.3
E	5.7	4.0	5.2	5.0
F	3.3	2.5	2.8	2.6
G	1.3	1.7	1.4	1.6
H	5.8	4.8	4.2	5.5
I	2.8	2.4	2.0	2.1
J	6.7	5.2	5.3	6.0
K	1.5	1.2	1.1	1.2
L	2.1	1.8	1.6	1.8
M	4.6	3.4	4.1	3.9
N	7.6	6.0	6.3	6.5
O	2.5	2.2	1.6	2.0

Table 1.2 shows measurements with different properties from those discussed so far. The first column gives the lengths of 15 pieces of string measured using a rule marked with divisions $\frac{1}{10}$th inch apart. The other three columns give the lengths of these 15 pieces of string as estimated by three subjects without the use of a rule. If the first column is assumed for the time being to be a true length (that is, measured without error) then a plot of the second column (Graham's estimates) against the first, for example, will give an indication of the measurement errors for the second column (see Figure 1.1). A similar example to this is given in Jaech (1985).

Looking at Figure 1.1, it can be seen that the slope of a best fitting straight line drawn through these points does not have a value of 1. It is lower than 1.

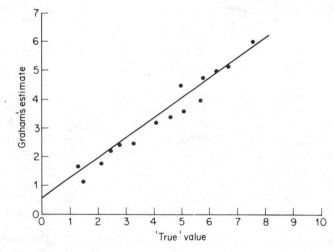

Fig. 1.1   Association between Graham's estimate of string lengths and their 'true' values (see Table 1.2). The line plotted is that obtained from a simple least squares regression.

The lengths of the pieces of string are consistently being underestimated and also the bias appears to be increasing as the true length of the string increases. This is understandable; the subject is likely to be able to guess the lengths of shorter lengths of string more accurately.

Let X_{ik} be the length of the ith piece of string as estimated by the kth subject ($i = 1, 2, \ldots, 15$; $k = 1, 2, 3$). Let μ_i represent the true length of the ith piece of string. A general model that can be used to relate X_{ik} to μ_i is

$$X_{ik} = \alpha_k + \beta_k \mu_i + e_{ik} \qquad (1.13)$$

As before e_{ik} represents a random error term and the es are assumed to be independent of each other and of the model parameters. The parameters α_k and β_k jointly describe the measurement bias characteristic of subject k. If $\beta_k \neq 1$ this indicates that bias is changing with μ_i; that is, in this measurement setting, bias is not constant. α_k and β_k are equivalent to the intercept and slope of the straight line relating the kth subject's length estimates with those provided by the use of a ruler.

1.3 Psychometric tests: the common-factor model

The three subjective judgements of lengths of pieces of string given in Table 1.2 are highly correlated. These correlations result from the measurement model described by Equation (1.13). In this example one can be reasonably confident that the measurement model is valid because the 'true' measured lengths of the 15 pieces of string are also available. In the measurement of psychological characteristics, however, one does not have access to the true scores. The measurement model is inferred from the pattern of correlations (or covariances) between fallible measurements. The measurement model is introduced to 'explain' or account for the correlations observed. The measurement model that is the most widely known in this area is the linear factor model originally proposed by Spearman in 1904 (see Lawley and Maxwell, 1971; Bartholomew, 1987).

Consider a battery of cognitive tests that can be used to measure intellectual ability. These could, for example, be examinations in school subjects such as mathematics, classics, English and French (Spearman, 1904) or the subtests of a modern intelligence test (Wechsler, 1981). Let Y_{ik} represent the score obtained by the ith subject on the kth cognitive test ($i = 1, 2, \ldots, I$; $k = 1, 2, \ldots, K$). Spearman's common-factor model postulates that each test score is made up of two components in the following way:

$$Y_{ik} = \lambda_k f_i + u_{ik} \qquad (1.14)$$

where the Y_{ik} are assumed to have been standardized to have zero mean and unit variance. The f_i term is a hypothetical true score for the ith individual on an underlying latent scale of cognitive ability (the common factor). The parameter λ_k is the constant of proportionality for test k and is referred to as the factor loading for test k. The last term, u_{ik}, is a measure for subject i that is specific to test k. For the case where each test is administered once only to each of the subjects the specific factors of the common-factor model are indistinguishable from measurement errors. As in the case of physical measurement it is usually assumed that the errors (specific factor scores) for different tests on the different individuals are uncorrelated with each other and with any of the common factor scores. Since the size of the loadings is dependent on the scale of measurement of the common factor an arbitrary constraint on the factor score is introduced. Usually the common-factor scores are constrained to have zero mean and unit variance.

From Equation (1.14) we have

$$\text{Var}(Y_{ik}) = \lambda_k^2 \text{Var}(f_i) + \text{Var}(u_{ik}) + 2\lambda_k \text{Cov}(f_i, u_{ik})$$

$$= \lambda_k^2 + \psi_k \tag{1.15}$$

where ψ_k is the variance of the specific factor for test k. (See Appendix 2 for the derivation of variances and covariances of linear combinations of random variables.) If the test scores, the Y_{ik}, have been standardized to have zero mean and unit variance, then

$$1 = \lambda_k^2 + \psi_k$$

or

$$\lambda_k^2 = 1 - \psi_k \tag{1.16}$$

Here λ_k^2 is the proportion of the variance of the kth test that is explained by the underlying common factor. The term *communality* is used to refer to this proportion.

For two standardized test scores, Y_{ik} and Y_{im},

$$\text{Cov}(Y_{ik}, Y_{im}) = \text{Corr}(Y_{ik}, Y_{im})$$

$$= E(Y_{ik}, Y_{im})$$

$$= E[(\lambda_k f_i + u_{ik})(\lambda_m f_i + u_{im})]$$

$$= \lambda_k \lambda_m \tag{1.17}$$

To summarize, the expected proportion of a test's variance that is explained by the common factor is the square of the test's loading (λ_k^2). The expected correlation between scores on two different tests is given by the product of their respective test loadings (that is, $\lambda_k \lambda_m$).

Returning to Equation (1.14), consider an alternative parameterization without standardization of the raw measurements or the common factor scores. Let the model now be

$$X_{ik} = \alpha_k + \beta_k \mu_i + e_{ik} \tag{1.18}$$

where X_{ik} is the measured score on test k for subject i, and α_k and β_k are parameters indicating constant and relative bias, respectively. The error term, e_{ik}, is assumed to be normally distributed with variance σ_k^2 and $\text{Cov}(e_{ik}, \mu_i) = 0$. The expected values of the observed scores are given by

$$E(X_{ik}) = \alpha_k + \beta_k \mu_i \tag{1.19}$$

and their variances by

$$\text{Var}(X_{ik}) = \beta_k^2 \sigma^2 + \sigma_k^2 \tag{1.20}$$

where σ^2 is the variance of the μ_i. Finally,

$$\text{Cov}(X_{ik}, X_{im}) = \beta_k \beta_m \sigma^2 \tag{1.21}$$

The similarity between Equations (1.13) and (1.18) is obvious. The only difference is that in Equation (1.13) the estimates by the three subjects are being regressed on a known true score whereas in Equation (1.18) the test scores are being regressed on a hypothetical latent trait. If the first column of Table 1.2 were to be lost or it were also assumed to be subject to measurement error then one would be led to the same common-factor model.

Tests that satisfy the measurement model described by Equation (1.18) are referred to as *congeneric* tests (Jöreskog, 1971a). The true score of test *m* $(\alpha_m + \beta_m \mu_i)$ is perfectly correlated with the true score of test *p* $(\alpha_p + \beta_p \mu_i)$. If the β_k of any two congeneric tests are equal, then those two tests are essentially *tau-equivalent*. They are tau-equivalent in the strict sense if the β_k are identical *and* the α_k are the same. Two tests are *parallel* if they are tau-equivalent *and* the variances of their measurement errors are also equal.

1.4 Poisson process models

So far the discussion has centred on measurement models that imply that the standard error of measurement is constant; that is, it does not depend on the characteristic of the object or subject being measured. This is an unduly restrictive assumption and there are many measurement settings in the physical, behavioural or social sciences where it is clearly not applicable (see below). In this section measurements assumed to be distributed following the Poisson distribution (Bulmer, 1979) will be described. Binary and categorical characteristics will be discussed in the following two sections, respectively.

In the physical sciences the archetypal Poisson process is radioactive decay. In a given sample of radioactive material that can be assumed to contain a single radioactive element such as tritium, for example, each atom will have a small fixed probability of undergoing a radioactive disintegration in a given time interval (which is also assumed to be fairly short). The probability of radioactive decay is the same for all of the tritium atoms in the sample and each atom decays completely independently of any other. If one now measures the number of radioactive decays in say 1 minute in a liquid scintillation counter, then the resulting count will be a Poisson variate. If the count obtained for radioactive sample *i* is X_i, then the following probability distribution can be assumed to apply:

$$P(X_i = x_i) = \frac{e^{-\tau_i}\tau_i^{x_i}}{x_i!} ; \qquad x_i \geq 0$$

It is a well-known characteristic of the Poisson distribution that the expected value of X_i is τ_i and so is the variance of X_i (see Exercise A1.1). The standard error of measurement of X_i is therefore $\sqrt{\tau_i}$. The parameter τ_i can be thought of as the true rate of radioactive decay for the *i*th sample.

Examples of time-dependent processes in the medical and social sciences are the numbers of births, deaths, marriages or divorces. One may wish to compare different time epochs or different human populations, or both. One well-known example occurs in the study of accidents. An owner of a bus fleet, for example, might wish to estimate the accident rate for different groups of drivers in several non-overlapping areas of a large city where the bus fleet operates. If the drivers in each area are thought to be independent of each other and to have a fixed but small risk of having an accident in a given time interval (so that the probability of any given driver having more than one accident is negligible) then a count of the number of drivers who have accidents in any one area of the city will be distributed according to Poisson distribution. The assumption that each driver has the same risk of an accident as any other may be untenable, of course.

Another well-known example of a measurement known to be a Poisson variate is in the estimation of blood cell or bacterial densities. A diluted sample of blood can yield a red cell count through the use of a traditional haemocytometer or a more modern electronic particle counter. Bacterial densities can similarly be estimated by the use of an electronic particle counter or a diluted sample can be evenly spread on a nutrient-containing surface and then incubated under sterile conditions to produce counts of visible colonies growing on this surface (each colony having arisen from an individual bacterium in the diluted culture sample).

Poisson models for certain types of psychometric tests have been proposed by Rasch (1960). One of these tests is designed to measure a particular aspect of an examinee's reading ability. The subject is presented with a text which he or she is required to read aloud, and a count of the number of words that are misread is recorded. It is assumed that a subject's probability of misreading any word in the text is a small constant characteristic of that subject but assumed not to depend on the particular word. It is also assumed that these probabilities are independent over words for a given person.

Let the probability of subject i misreading a word in oral reading test k be P_{ik}. Further, let this probability be the ratio of two parameters – the test's difficulty, β_k, and the subject's reading ability, μ_i. That is

$$P_{ik} = \frac{\beta_k}{\mu_i} \tag{1.22}$$

Clearly the parameter values (that is, the values β_k and μ_i) have to be constrained so that P_{ik} has a value between 0 and 1. If X_{ik} is the number of misreadings made by subject i on reading test k, then

$$X_{ik} = N_k P_{ik} + e_{ik} \tag{1.23}$$

and

$$E(X_{ik}) = N_k P_{ik} = N_k \beta_k / \mu_i \tag{1.24}$$

where N_k is the total number of words to be read from test k. If $E(X_{ik})$ is represented by τ_{ik}, then

$$P(X_{ik} = x_{ik} \mid \tau_{ik}) = \frac{e^{-\tau_{ik}} \tau_{ik}^{x_{ik}}}{x_{ik}!} \; ;$$
$$x_{ik} = 0, 1, 2, \ldots \tag{1.25}$$

and, from Equation (1.24),

$$\log_e(\tau_{ik}) = \log_e N_k + \log_e \beta_k - \log_e \mu_i \tag{1.26}$$

This is an example of a simple log-linear model (see Bishop *et al.*, 1975). It is called a log-linear model because it is an additive or linear model ($\log N_k + \log_e \beta_k - \log_e \mu_i$) for a logarithm of the expected value, τ_{ik}.

Feldstein and Davis (1984) discuss a similar model to that of Rasch for a rater reliability problem. Their practical example is a table of counts of the number of different kinds of avoiding strategies observed by nine raters independently judging five 15-minute videotaped interviews. Let X_{ik} represent the number of avoiding strategies observed in videotape i by rater k, and let the expected value of X_{ik} be τ_{ik}. Suppose that the distribution of avoiding strategies for subject i is Poisson with mean μ_i. Also suppose that a rater will not mistakenly record an avoiding strategy when it is absent, but if the behaviour occurs the probability that rater k will detect it is β_k. If β_k is not dependent on the videotape being rated, but is a fixed characteristic of rater k, then $E(X_{ik}) = \tau_{ik} = \beta_k \mu_i$. The probability that rater k detects exactly x_{ik} avoiding strategies in tape i is again given by

$$P(X_{ik} = x_{ik} \mid \tau_{ik}) = \frac{e^{-\beta_k \mu_i}(\beta_k \mu_i)^{x_{ik}}}{x_{ik}!} \; ;$$

$$x_{ik} = 0, 1, 2, \ldots$$

$$(1.27)$$

and

$$\log_e(\tau_{ik}) = \log_e \beta_k + \log_e \mu_i \qquad (1.28)$$

This, too, is a simple log-linear model. It differs from (1.26) in only having two terms to the right-hand side because the videotape length is constant in this example and has been incorporated into the β_k parameters. However, if the video recordings were to differ in length, with videotape i having length L_i, then the appropriate model could be

$$\log_e(\tau_{ik}) = \log_e L_i + \log_e \beta_k + \log_e \mu_i \qquad (1.29)$$

Feldstein and Davis (1984) also adapted their multiplicative model to allow raters to mistakenly record behaviours when they were in fact absent. That is, the model now allows for false positives as well as false negatives. If another parameter, α_k, is introduced to allow for a fixed personal bias for rater k, then the new model is

$$E(X_{ik}) = \tau_{ik} = \alpha_k + \beta_k \mu_i \qquad (1.30)$$

Note that this is no longer equivalent to a log-linear model but it does have the same form as those described by Equations (1.13), (1.14) and (1.18); the only difference being an error structure characteristic of the Poisson rather than the normal distribution.

1.5 Binary response models

Consider a cognitive test, such as an examination of mathematical ability, and let the test contain N items or questions of equal difficulty. That is, for examinee i the probability of getting item k correct is simply p_i (for all k). Further assume that an examinee's performance on item k is independent of that for item m (for all m and k where $k \neq m$). The total number of correct answers is used as a measure of the subject's mathematical ability. If X_i is the number-right score for subject i, then

$$X_i = Np_i + e_i \qquad (1.31)$$

and

$$E(x_i) = \tau_i = Np_i \tag{1.32}$$

Each item can be considered as an independent Bernoulli trial and therefore the number-right score is distributed as a binomial variate. That is,

$$P(X_i = x_i) = \binom{x_i}{N} p_i^{x_i}(1 - p_i)^{N-x_i} \tag{1.33}$$

It follows that the number-right score has variance given by

$$\sigma_{X_i}^2 = Np_i(1 - p_i) \tag{1.34}$$

The variance of the number-right score is not constant nor is it independent of the subject's ability τ_i. It approaches zero as p_i approaches either zero or unity, and is at a maximum when $p_i = \frac{1}{2}$.

Consider a possible linear model for p_i. Because of the problems caused by the restricted range of the p_i (0–1), it is usual to model some transformation of p_i (see Cox, 1970). First however, the assumption of constant item difficulty will be relaxed, but items will still be assumed to be locally independent (a subject's performance on one item is independent of his or her performance on another). Let the probability of subject i getting item k correct be p_{ik}. The odds on the subject getting the item correct is $p_{ik}/(1 - p_{ik})$. The odds can range from zero to infinity. Following Equation (1.22) let the odds be a ratio of two parameters. That is,

$$\frac{p_{ik}}{1 - p_{ik}} = \frac{\mu_i}{\beta_k} \tag{1.35}$$

where, as before, β_k is a measure of the difficulty of item k and μ_i is a measure of the subject's ability. Taking logarithms on both sides of Equation (1.35) gives

$$\log_e\left(\frac{p_{ik}}{1 - p_{ik}}\right) = \log_e \mu_i - \log_e \beta_k \tag{1.36}$$

$$= m_i - b_k \tag{1.37}$$

where $m_i = \log_e \mu_i$ and $b_k = \log_e \beta_k$. This is equivalent to

$$p_{ik} = \frac{e^{(m_i - b_k)}}{1 + e^{(m_i - b_k)}} \tag{1.38}$$

or

$$p_{ik} = \frac{1}{1 + e^{-(m_i - b_k)}} \tag{1.39}$$

Equation (1.36) is an example of a linear-logistic model for a binary response, k, the term on the left being the logit of the proportion p_{ik}. The model is the simplest form of several different latent trait models used in psychometric test theory (see Lord and Novick, 1968 and Lord, 1980). It is usually known as the Rasch model (Rasch, 1960) but should be distinguished from the Rasch model introduced in Section 1.4. A linear-logistic model for binary responses to survey questionnaire items has been discussed by Anderson and Aitken (1985).

1.6 Identification and diagnosis

This section is concerned with measurements made on a nominal or categorical scale. The categories may or may not be ordered depending on the meaning placed on the categories by the observer, and, of course, a special case of categorical measurement that has already been discussed is a simple binary response. A subject may be sick or well, an alcoholic or not an alcoholic, and so on. In psychiatry, as well as in general medicine, a clinician is usually interested in placing a patient in a particular diagnostic category (depressed, anxious, schizophrenic, and so on) prior to choosing a method of treatment. An example of an ordered series of categories is social class. Subjective rating scales are other examples of ordinal categorical responses. There is also the possibility of complex descriptions of subjects; patients may, for example, be described by a profile of categorical or ordinal responses or by multiple diagnoses.

There appears to be no simple measurement model applicable to this wide variety of settings. One is usually interested in assessing patterns of agreement and disagreement between raters or ratings and, as in the case of interval and ratio measurements, one can search for systematic differences between raters (bias) as well as for haphazard or random disagreements. A detailed discussion of this problem will be left to the following chapter.

1.7 Comparison of measurements with a standard

Suppose one has a well-established accurate method of measurement of the concentration of a given chemical compound, but that it is too expensive and cumbersome for routine use. A cheap and easy alternative method is developed which is known to be imprecise and possibly biased. The two methods can be compared through the use of a *calibration* experiment. Alternatively, a calibration experiment need not involve the use of an accurate standard measuring instrument. One could, for example, make up a series of solutions with a known concentration of the compound of interest and then attempt to measure the concentration of compound in each of the solutions using the new assay method. A third example is the bioassay. Here a biological response is elicited by each of a series of dilutions of a drug or active compound with an unknown concentration. These biological responses are then compared with responses elicited by a series of similar dilutions of a standard drug preparation of known concentration. By comparison of the two dose–response curves the potency of the unknown drug preparation relative to the standard can be estimated.

Figure 1.1 is an example of a simple linear calibration curve. The string length as guessed by the subject is a quick and easy measurement. The length measured by a rule can be regarded as the string's true length. From a practical point of view this is obviously a rather frivolous example, but it is an analogue of calibration curves that are routinely produced in analytical chemistry laboratories.

Ignoring the other subject's guesses of the string lengths, the calibration model for Figure 1.1 is

$$X_i = \alpha + \beta\mu_i + e_i \tag{1.40}$$

where X_i is the measured (guessed) length of string i and μ_i is the corresponding true length. The parameters α and β (intercept and slope) refer to constant bias and relative bias, respectively. A computer program such as GLIM (see Appendix 4) can be used to fit the straight line $X_i = \alpha + \beta\mu_i$ to the data in the first two columns of Table 1.2. The least squares estimates of α and β are 0.32 (standard error 0.19) and 0.72 (standard error 0.04), respectively. The residual mean square, the estimate of the variance of the error terms σ_e^2, is 0.09. These results indicate that $\hat{\alpha}$ is not significantly different from zero. If α is now constrained to be equal to zero (that is, the model is now $X_i = \beta\mu_i + e_i$), then the least squares estimates of β and σ_e^2 are 0.79 (standard error 0.18) and 0.10, respectively. The parameter β could also have been constrained to equal 1 instead (through the use of the OFFSET directive in GLIM), but this constraint is clearly unrealistic in the light of the present data. Table 1.3 presents the sums of the squared residuals after fitting various calibration models using GLIM. It will be left as an exercise for the reader to construct appropriate F-tests to justify the above conclusions.

Table 1.3 Linear regression of Graham's estimates (X) on true string lengths (μ)

Fitted model	Residual S.S.	Residual d.f.
$X = \alpha$	29.69	14
$X = \alpha + \beta\mu$	1.17	13
$X = \beta\mu$	1.42	14
$X = \alpha + \mu$	5.28	14
$X = \mu$	16.37	15

Now, the main purpose of a calibration model is to subsequently use the parameter estimates, together with the estimate of σ_e^2, to estimate the value of an unknown μ_* from a given measurement X_*. In the present example μ_* is estimated from

$$\hat{\mu}_* = \frac{X_*}{\hat{\beta}} \tag{1.41}$$

So, if $X_* = 3.0$, for example, then $\hat{\mu}_* = 3.0/0.79 = 3.8$. Now suppose that a confidence interval for μ_* is required.

The variance of $\hat{\mu}_*$ could be estimated using the approximate methods described in Appendix 3 (the delta technique). Here, however, a more exact method of obtaining the required confidence interval will be illustrated. This is based on the use of Fieller's theorem (see Colquhoun, 1971, Chapter 13). Consider the linear combination $X_* - \hat{\beta}\mu_*$ with X_* and $\hat{\beta}$ assumed to be normally distributed random variables and μ_* fixed. Then $E(X_* - \hat{\beta}\mu_*) = 0$ and

$$\mathrm{Var}(X_* - \hat{\beta}\mu_*) = \mathrm{Var}(X_*) - 2\mu_*\,\mathrm{Cov}(\hat{\beta}, X_*) + \mu_*^2\,\mathrm{Var}(\hat{\beta}) \tag{1.42}$$

In this case, since X_* and $\hat{\beta}$ are independent,

$$\mathrm{Var}(X_* - \hat{\beta}\mu_*) = \mathrm{Var}(X_*) + \mu_*^2\,\mathrm{Var}(\hat{\beta}) \tag{1.43}$$

For fixed μ_*, $\mathrm{Var}(X_*) = \sigma_e^2$, and, finally,

$$\mathrm{Var}(X_* - \hat{\beta}\mu_*) = \sigma_e^2 + \mu_*^2\,\mathrm{Var}(\hat{\beta}) \tag{1.44}$$

See Appendix 2 for the derivation of the variance of linear combinations of random variables.

Now, as we have assumed that both X_* and $\hat{\beta}$ are normally distributed, so the combination $(X_* - \hat{\beta}\mu_*)$ will also be normally distributed. The ratio

$$\frac{(X_* - \hat{\beta}\mu_*)}{\text{s.d. } (X_* - \hat{\beta}\mu_*)}$$

will be distributed as a t-statistic with 14 degrees of freedom if the standard deviation of $(X_* - \hat{\beta}\mu_*)$ is estimated from the analysis of the data plotted in Figure 1.1.

A 95% confidence interval for $(X_* - \hat{\beta}\mu_*)$ can be obtained from the inequality

$$\frac{(X_* - \hat{\beta}\mu_*)}{\hat{\sigma}_e^2 + \mu_*^2 \, \text{Vâr}(\hat{\beta})} \leq 2.15^2 = 4.62$$

where 2.15 is the appropriate critical value for t with 14 degrees of freedom. The boundaries of this confidence interval can be obtained by replacing the inequality by an equality and then solving the quadratic equation

$$(X_* - \hat{\beta}\mu_*)^2 = 4.62\hat{\sigma}_e^2 + 4.62\mu_*^2 \, \text{Vâr}(\hat{\beta})$$

This will be left as an exercise for the reader (note that it will result in an asymmetric interval about the estimated value, $\hat{\mu}_* = 3.9$).

In psychiatric research use is often made of screening questionnaires that are completed by the subject without the help of a psychiatrist or other expert. A well-known example is the General Health Questionnaire or GHQ (Goldberg, 1972). Clearly the GHQ is a fallible measuring instrument, and if it is to be used as a cost-effective method of assessing the prevalence of psychiatric disorders in different sections of the community it is vital that its 'validity' is first checked using representative samples from that community. Validation in this context is just another form of calibration. The infallible measure is usually taken to be the result of a thorough psychiatric interview. A group of subjects who after interview are found to be true 'cases' (that is, suffering from a sort of psychiatric distress) is compared with a similar group of patients found not to be 'cases' on interview. The two groups are given the GHQ to complete (in practice the GHQ is usually completed prior to the full interview since the interview influences the way in which the GHQ is completed). On the basis of his or her answers to the items on the GHQ, a subject will be given a label of 'GHQ + ve' (that is, a 'case') or 'GHQ − ve'. The results of this validation exercise can be summarized by the following simple table of proportions.

	GHQ + ve	GHQ − ve	Total
Case	$1 - \beta$	β	1
Non-case	α	$1 - \alpha$	1

Note that the proportions α and β are both *row* proportions. β is the proportion of true cases that are GHQ − ve; that is, the proportion of *false negatives*. The proportion $(1 - \beta)$ is the *sensitivity* of the screening test. α is the proportion of non-cases that are GHQ + ve; that is, the proportion of *false positives*. The proportion $(1 - \alpha)$ is the *specificity* of the test. If the true proportion of cases (prevalence) is π and the observed or fallible prevalence estimated for the proportion of GHQ + ves in a sample is p, then the appropriate measurement model is

$$p = \pi(1 - \beta) + (1 - \pi)\alpha \tag{1.45}$$

For an observation $\hat{p}$, the appropriate prevalence estimate is given by

$$\hat{\pi} = \frac{\hat{p} - \alpha}{1 - \beta - \alpha} \tag{1.46}$$

$$= \frac{\hat{p}}{1 - \beta - \alpha} - \frac{\alpha}{1 - \beta - \alpha} \tag{1.47}$$

On the assumption that $\hat{p}$, α and β are all binomial, the standard error for $\hat{\pi}$ could be estimated using the methods described in Appendix 3.

1.8   Implications of measurement error

The social and economic costs of measurement or diagnostic errors should be obvious. A woman at risk for cancer of the cervix does not want to be subjected to a screening test that is likely to miss a malignancy if present; nor does she want to face the anxiety and the other possible consequences of being a false positive. Similarly if a psychometric test is being used as part of a selection process for a job or college entrance it needs to have both high reliability and high validity. Returning to a medical example, it is essential that a pathologist is able to distinguish different types of malignancy from biopsy samples, particularly if they have different prognostic implications for the patient. In the field of industrial quality control it is clearly desirable to be able to distinguish product variability from measurement errors. One does not wish to let many faulty items through to the market place nor is it desirable to reject items as faulty that in reality are perfectly acceptable to the customer. Finally, in fields such as market research one wishes to be able to detect and cope with problems caused by interviewer bias as well as the more obvious sampling biases.

The influence of measurement errors on scientific research may not be quite so obvious. They clearly lead to lack of precision in simple measurements and comparisons and also lead to lower power for statistical significance tests. They also, however, lead to potential biases; and the source of these biases may not be at all obvious to a statistically naive research worker. To illustrate this point two examples will be discussed. The first involves the use of a medical screening questionnaire used to assess the prevalence of psychiatric problems in men and women. The aim of the study might be to compare prevalence levels in the two sexes. The second example involves the use of analysis of covariance to assess the difference between two groups after allowing for a covariate that is subject to measurement error.

The calibration of a screening test such as the GHQ was discussed in Section 1.8. Let P_m and P_F be the true proportions of GHQ + ves in the male and female populations, respectively. Let the sensitivity of the GHQ for males be $(1 - \beta_m)$ and that for females be $(1 - \beta_F)$. Similarly let the corresponding specificities for males and females be $(1 - \alpha_m)$ and $(1 - \alpha_F)$, respectively. Finally let the true prevalence of psychiatric disorder in the two sexes be π_m and π_F.

From Equation (1.45) it follows that

$$P_m = \pi_m(1 - \beta_m) + (1 - \pi_m)\alpha_m \tag{1.48}$$

and

$$P_F = \pi_F(1 - \beta_F) + (1 - \pi_F)\alpha_F \tag{1.49}$$

The expected sex difference in psychiatric morbidity, as measured by the fallible screening questionnaire, is given by

$$(P_m - P_F) = \pi_m(1 - \beta_m) + (1 - \pi_m)\alpha_m - \pi_F(1 - \beta_F) - (1 - \pi_F)\alpha_F \tag{1.50}$$

In general the equality of π_m and π_F will not imply that $P_m = P_F$. If it is safe to assume that $\beta_m = \beta_F$ and that $\alpha_F = \alpha_m$, then (1.50) can be simplified to yield the following expression.

$$(P_m - P_F) = \pi_m(1 - \beta_m) + (1 - \pi_m)\alpha - \pi_F(1 - \beta) - (1 - \pi_F)\alpha$$

$$= (\pi_m - \pi_F)(1 - \beta - \alpha) \tag{1.51}$$

In general $(P_m - P_F)$ will underestimate the value of $(\pi_m - \pi_F)$; that is, $(P_m - P_F)$ will provide biased estimates of the difference in morbidity in two populations. If β and α are large enough to make $(\alpha + \beta) > 1$, then the direction of the difference will be the reverse of the observation $(P_m - P_F)$!

Now consider a second example: the use of the analysis of covariance when the covariate is measured subject to error. Suppose a psychologist carries out an observational study (or quasi-experiment) involving two groups of subjects. Group A receives one form of training and group B another. At the end of the experiment the investigator measures the subject's skill at carrying out the appropriate task. The outcome score for individual i is represented by y_i and the mean outcomes for groups A and B are $\bar{y}_A$ and $\bar{y}_B$, respectively. The investigator realizes that the subjects' skill at the beginning of the study influences the outcome and she decides to use a measure of this skill, denoted by x_i for the ith subject, as a covariate in an analysis of covariance; the aim being to decrease the bias that may have arisen because of difference in initial ability.

The initial score, x_i, is measured subject to error. Let

$$x_i = \mu_{x_i} + \delta_i \tag{1.52}$$

and

$$y_i = \mu_{y_i} + \varepsilon_i \tag{1.53}$$

where $\mu_{x_i} = E(x_i)$ and $\mu_{y_i} = E(y_i)$.

Fitting a straight line

$$E(y_i) = \alpha + \beta_{x_i} \tag{1.54}$$

to the subjects' scores yields the following estimate of the slope, β.

$$\beta' = \frac{\text{Cov}(x, y)}{\text{Var}(x)} \tag{1.55}$$

$$= \frac{\text{Cov}(\mu_x, y) + \text{Cov}(\delta, y)}{\text{Var}(\mu_x) + \text{Var}(\delta) + 2\,\text{Cov}(\mu_x, \delta)} \tag{1.56}$$

$$= \frac{\text{Cov}(\mu_x, y)}{\text{Var}(\mu_x) + \text{Var}(\delta)}$$

(assuming that $\mathrm{Cov}(\delta, y) = \mathrm{Cov}(\mu_x, \delta) = 0$). The true regression slope, however, should be estimated for

$$\hat{\beta} = \frac{\mathrm{Cov}(\mu_x, y)}{\mathrm{Var}(\mu_x)} \tag{1.57}$$

From Equations (1.56) and (1.57)

$$\begin{aligned}
\hat{\beta}' &= \frac{\mathrm{Var}(\mu_x)}{\mathrm{Var}(\mu_x) + \mathrm{Var}(\delta)} \hat{\beta} \\
&= \frac{\mathrm{Var}(\mu_x)}{\mathrm{Var}(x)} \hat{\beta}
\end{aligned} \tag{1.58}$$

The observed slope, $\hat{\beta}'$, is clearly biased. On average it will underestimate the slope, $\hat{\beta}$.

If the investigator now uses the estimate $\hat{\beta}$ to produce an estimate of the group differences adjusted for the covariate x, then the following is obtained. Adjusted difference, say, d'_{AB} is given by

$$d'_{AB} = \bar{y}_A - \bar{y}_B - \hat{\beta}'(\bar{x}_A - \bar{x}_B) \tag{1.59}$$

If x were not subject to measurement error, the difference would here be given by

$$d_{AB} = \bar{y}_A - \bar{y}_B - \hat{\beta}(\bar{x}_A - \bar{x}_B) \tag{1.60}$$

The latter expression yields unbiased estimates of the true treatment difference whereas Equation (1.59) will not. The bias will be reduced by the covariance analysis using covariates subject to measurement error, but not completely. It is even possible for d'_{AB} and d_{AB} to have different signs.

1.9 Exercises

1.1 Analyse the difference scores obtained from Table 1.1 and estimate the standard error measurement of the scales. Compare your results with those of the simpler experiments described at the beginning of the chapter. Comment on the mean difference of the weights at the two times.

1.2 Use a linear regression program such as GLIM to compare Brian's estimates in Table 1.2 with the true values. Considering the regression line as an example of a calibration curve, calculate a 95% confidence interval for a true length given an estimate of 5.0 inches by Brian.

1.3 Fit a linear regression line to David's estimates in Table 1.2 and compare the parameter estimates obtained with those obtained by fitting a straight line to those of Graham and Brian.

1.4 This chapter has introduced three methods of studying the performance of some kitchen scales. Discuss the relative merits of these approaches together with any other ways of investigating measurement errors that you might find useful.

1.5 Comment on likely problems of the use of repeated measurement to assess the errors in psychometric test scores.

2

Measures of Proximity, Agreement and Disagreement

2.1 Introduction

In the first chapter the discussion concentrated on measurement models with stress being placed on concepts of bias and precision. In the present chapter the complementary idea of the proximity of ratings or raters will be developed. There will, on the whole, be no reference to a particular measurement model, nor will there be any mention of 'true' scores. Here one is simply interested in the closeness or similarity of ratings as a measure of both agreement and consistency (both within and between raters and ratings). Closely linked to the concepts of similarity and dissimilarity of ratings is the idea that the level of disagreement between any two ratings (or raters) might be represented by some sort of distance measure. In this context distance is not a property relating two points in a physical space but a measure in an abstract space that has the same mathematical properties as that used for the more familiar three-dimensional physical world.

A simple measure or index of agreement viewed in isolation may be difficult to interpret. It has to be considered in the context of the range of its possible values. This is, of course, dependent on the characteristics of the sample or population of objects or subjects being measured or rated. It is, or might be, also scale dependent. It would, for example, be a little disconcerting to find that an index of agreement between weighing machines were dependent on whether the weights were recorded in grammes or in ounces.

Another important goal is to produce a measure of agreement that represents an improvement in the similarity between any ratings over and above that which might have been expected by chance. Again, this will be dependent on the characteristics of the population of objects or subjects being studied. This goal, in turn, will very often lead to the derivation of a test to assess whether or not a particular level of agreement has any statistical significance (that is, could the level of agreement easily have occurred by chance?). A cynic, of course, might point out that this is a rather worthless exercise since one's aims should be a little more ambitious than being able to do better than the random allocation of measurement to subjects. On the other hand, considerations of statistical power should lead to a study of inter-rater agreement or of instrument precision that is capable of yielding useful information. Too often an investigation of this type is of negligible value because of the use of inadequate sample sizes.

2.2   Similarity of interval- and ratio-scaled measurements

Consider the weight of the first item given in Table 1.1. At the first weighing it was recorded as 560 g. At the second time it was 550 g. The difference between the two measurements is simply 10 g. As a measure of disagreement between these two recordings it is clearly desirable that the value obtained would have been the same if the first recording had been 550 g followed by a second recording of 560 g. In general, for the ith object in Table 1.1 one could use the absolute value of $x_{i1} - x_{i2}$ (that is, $|x_{i1} - x_{i2}|$) as the required disagreement index. This index could be standardized for the sample of objects under consideration by looking for the largest index (60 in the present example) and dividing the observed disagreement for the first item, for example, by this maximum disagreement index. Here the required value is 10/60 or 0.17. For any given sample the disagreement measures will now range from a minimum value of 0 to a maximum of 1. To get a measure of similarity or agreement between these two ratings one could simply subtract the disagreement index from 1. That is

$$d_{ii} = |x_{i1} - x_{i2}|/\max(|x_{i1} - x_{i2}|) \qquad (2.1)$$

$$s_{ii} = 1 - d_{ii} \qquad (2.2)$$

Returning to the first object in Table 1.1, the square of the difference between the two weights could have been chosen as the basic disagreement index instead of the absolute value. If an overall measure of disagreement is needed for Table 1.1 (that is, a summary statistic for all of the 20 disagreements) one could use either the sum or the arithmetic mean of the disagreement measures for each of the 20 items (with or without standardization of the individual disagreement indices). In the case of the squared values, $(x_{i1} - x_{i2})^2$, the arithmetic mean would have a clear interpretation in terms of the measurement of instrument precisions through the use of 'error' variances (as was discussed in the first chapter).

Now consider the concept of the *distance* between the sets of weights at time 1 and those at time 2. The most familiar distance measure is the *Euclidean* metric. In the present example this is defined by

$$D_{12} = \left[\sum_{i=1}^{20} (x_{i1} - x_{i2})^2 \right]^{1/2} \qquad (2.3)$$

If only two objects were being weighed this would simply be equivalent to the expression of Pythagoras' theorem as seen in elementary geometry texts. Here the distance between occasions is being measured in an abstract rather than physical space, and with 20 dimensions rather than the customary 2 or 3. Another commonly used metric is the *absolute* or *city-block* distance. This is given by

$$D_{12} = \sum_{i=1}^{20} |x_{i1} - x_{i2}| \qquad (2.4)$$

2.3 Similarity of binary responses

Consider Table 2.1. This shows the examination results of 29 candidates whose scripts are each marked by two examiners who categorize the examinees as Passing (P) or Failing (F) the tests. On the far right of the table is an indication of agreement: Yes is represented by Y; No by N. Using the same ideas for the measurement of similarity and distance as were introduced in the last section one can represent a disagreement between examiners by the distance value 1 and agreement by 0. The overall distance between the examiners is simply the sum of the disagreements (7) if one chooses the city-block metric or $\sqrt{7}$ if the Euclidean metric is chosen. The average distance between the examiners is 7/29 (number of disagreements/number of candidates).

Table 2.1 Agreement between two examiners on passing (P) or failing (F) an examination

Candidate	Examiner A	Examiner B	Agreement
1	F	P	N
2	F	F	Y
3	F	F	Y
4	P	P	Y
5	F	F	Y
6	P	P	Y
7	F	F	Y
8	F	F	Y
9	F	F	Y
10	P	P	Y
11	F	P	N
12	P	P	Y
13	F	F	Y
14	P	P	Y
15	P	P	Y
16	F	P	N
17	F	P	N
18	F	P	N
19	P	P	Y
20	F	F	Y
21	P	P	Y
22	P	P	Y
23	F	F	Y
24	F	F	Y
25	P	P	Y
26	F	P	N
27	F	P	N
28	P	P	Y
29	P	P	Y

Of course, one could have chosen to measure similarity or agreement rather than distance. An agreement could simply be scored as 1 and a disagreement as 0. The proportion of candidates where there is agreement is simply 22/29. This could be multiplied by 100 to get the corresponding percentage agreement. Note that if the proportion of agreement or similarity between Examiner A and Examiner B is S_{AB} and the Euclidean distance is D_{AB}, then

$$D_{AB} = (1 - S_{AB})^{1/2} \tag{2.5}$$

S_{AB} is equivalent to the simple matching coefficient of numerical taxonomy (Dunn and Everitt, 1982).

The observations in Table 2.1 could be summarized by a simple two-way contingency table as follows:

	Examiner B		
	F	P	Total
Examiner A F	10	7	17
P	0	12	12
Total	10	19	29

The simple matching coefficient or proportion agreement is given by

$$S_{AB} = \frac{10 + 12}{10 + 7 + 0 + 12}$$

$$= \frac{22}{29}, \text{ as before.} \tag{2.6}$$

Now consider a typical two-way table representing patterns of agreement and disagreement between two raters:

	Rater B		
	Yes	No	Total
Rater A Yes	a	b	$a + b$
No	c	d	$c + d$
Total	$a + c$	$b + d$	$a + b + c + d$

As before, the simple matching coefficient is

$$S_{AB} = \frac{a + d}{a + b + c + d} \tag{2.7}$$

But suppose that this is an example from epidemiology and the raters are diagnosing a rare condition. The fact that they agree on the absence of the condition (the count d) may be thought of as being rather uninformative. A better measure of agreement in this case would be the proportion of agreements when at least one of the raters thinks that the condition is present. One possible agreement measure is

$$S'_{AB} = \frac{a}{a + b + c} \tag{2.8}$$

This is the Jaccard coefficient of numerical taxonomy (Dunn and Everitt, 1982), and the corresponding Euclidean distance is given by

$$D'_{AB} = (1 - S'_{AB})^{1/2} \tag{2.9}$$

Other indices of agreement for binary responses can be found in Landis and Koch (1975) or in Fleiss (1975, 1981b).

2.4 Similarity of categorical responses

In the last section the agreement between two examiners was discussed. These two examiners actually scored the examination scripts as indicated in Table 2.2. These scores form an ordinal scale from 0, indicating a very poor performance to 4, indicating an excellent result. The difference between Examiner A and Examiner B for each of the 29 candidates is shown on the right of the table. If the signs of these differences are ignored one can simply add them to produce a city-block distance indicating disagreement between these two examiners ($\sum_{i=1}^{29} |d_{ii}| = 20$). Alternatively they can be squared and then added ($\sum_{i=1}^{29} d_{ii}^2 = 24$). The Euclidean distance between the raters is simply the square root of the latter sum. Now consider a 5×5 square contingency table produced for Table 2.2. This is shown in part (a) of Table 2.3. In part (b) of the same table is shown the absolute differences between all possible combinations of ratings by this pair of examiners. Following each entry in this part of the table there follows the corresponding squared difference (in brackets). The distance between Examiner A and Examiner B, taken over all candidates, is a weighted sum of the cell counts in Table 2.3(a) with weights provided by the entries in Table 2.3(b). These entries will be called *disagreement weights*.

Table 2.2 Agreement between examination marks for the candidates in Table 2.1

Candidate	Examiner A	Examiner B	Difference
1	1	2	−1
2	0	0	0
3	0	0	0
4	2	2	0
5	0	0	0
6	4	3	1
7	0	0	0
8	0	0	0
9	0	0	0
10	2	3	−1
11	1	2	−1
12	2	3	−1
13	0	1	−1
14	4	3	1
15	4	3	1
16	1	2	−1
17	0	2	−2
18	1	2	−1
19	2	3	−1
20	0	0	0
21	2	3	−1
22	4	4	0
23	0	0	0
24	0	0	0
25	4	3	1
26	0	2	−2
27	1	2	−1
28	3	4	−1
29	2	3	−1

Table 2.3 Summaries of the candidates' performance in Table 2.2

(a) *Counts*

| | | Rating by Examiner B | | | | | |
		0	1	2	3	4	Total
	0	9	1	2	0	0	12
Rating by	**1**	0	0	5	0	0	5
Examiner A	**2**	0	0	1	5	0	6
	3	0	0	0	0	1	1
	4	0	0	0	4	1	5
	Total	9	1	8	9	2	29

(b) *Absolute differences and squared differences (in brackets) between ratings*

| | | Rating by Examiner B | | | | |
		0	1	2	3	4
	0	0(0)	1(1)	2(4)	3(9)	4(16)
Rating by	**1**	1(1)	0(0)	1(1)	2(4)	3(9)
Examiner A	**2**	2(4)	1(1)	0(0)	1(1)	2(4)
	3	3(9)	2(4)	1(1)	0(0)	1(1)
	4	4(16)	3(9)	2(4)	1(1)	0(0)

Ignoring the zero counts in Table 2.3(a), the required city-block metric is given by

$$D_{AB} = (9 \times 0) + (1 \times 1) + (2 \times 2) + (5 \times 1) + (1 \times 0) + (5 \times 1)$$
$$+ (1 \times 1) + (4 \times 1) + (1 \times 0)$$
$$= 20$$

The square of the corresponding Euclidean distance is given by

$$D_{AB}^2 = (9 \times 0) + (1 \times 1) + (2 \times 4) + (5 \times 1) + (1 \times 0) + (5 \times 1)$$
$$+ (1 \times 1) + (4 \times 1) + (1 \times 0)$$
$$= 24$$

The absolute differences given in Table 2.3(b) are often referred to as *linear disagreement weights*; the squared differences are referred to as *quadratic disagreement weights*. A much cruder form of weighting could have simply involved the use of 0s on the diagonal entries of Table 2.3(b) (indicating agreement) and 1s in the off-diagonal cells (indicating disagreement). The city-block metric in this case would be reduced to the total number of disagreements.

Note that the choice of disagreement weights is made by the investigator. Other weights could have been used than the simple linear and quadratic ones described. If the response categories cannot be ordered, as in the case of medical diagnoses, for example, then the distance between the individual diagnostic categories is dependent on clinicians' perceptions of their similarities. Again the distance between all pairs of the different diagnostic categories could be fixed at 1, but this might not lead to an optimal use of the information provided by clinical assessments. As an example of the use of subjective distances or weights, consider a diagnostic exercise were psychiatrists to have to place each of several patients in one of three mutually exclusive categories: schizophrenia, psychotic depression and neurotic depression. If one psychiatrist gives a diagnosis of schizophrenia to a patient whilst another gives a diagnosis of psychotic depression, one could decide that the distance between these two psychiatric diagnoses is 1. If the second psychiatrist had given the diagnosis of neurotic depression this might have been seen as an indicator of greater disagreement and here the distance might be recorded as 2. Similarly the distance between psychotic depression and neurotic depression might be chosen to be either 1 or 2. Setting the latter distance to 1 while keeping the distance between schizophrenia and psychotic depression as 1 together with a distance between schizophrenia and neurotic depression of 2 is equivalent to accepting that the sequence schizophrenia, psychotic depression, neurotic depression forms some sort of ordinal scale. The choice of distances (weights) can be made to imply complete ordering of categories, or only partial ordering, or no ordering at all. In each of three situations the relative sizes of the different distances or weights can still be used to convey information on the proximity of the different pairs of diagnostic categories. The weights could be chosen to express the results of a hierarchical classification of labels or diagnostic categories. Consider, for example, the following diagnostic codes:

A anxiety neurosis
B depressive neurosis
C insomnia
D tension headache
E schizophrenia
F affective psychosis
G alcoholism and drug dependence.

Now suppose that a clinician has classified these disorders on the basis of their perceived similarities into the following hierarchy:

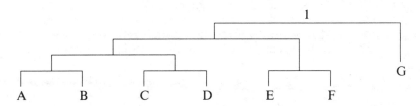

This hierarchy could be reflected by several choices of disagreement weights. One possibility is shown in Table 2.4. Any monotonic transformation (i.e. one that maintains the same rank order) of the numbers in Table 2.4 would, of course, convey the same information concerning the structure of the hierarchy, but would convey different perceptions about the distances within and between groups of the psychiatric diagnoses.

2.5 Similarity of response profiles

Quite often an individual is not given a simple or diagnostic label. A subject may be rated for several different (but possibly related) symptoms or characteristics. A patient may be asked to complete a battery of several psychometric

Table 2.4 Hypothetical disagreement weights for psychiatric diagnoses

		Diagnosis *						
		A	B	C	D	E	F	G
	A	0	1	2	2	3	3	4
	B	1	0	2	2	3	3	4
	C	2	2	0	1	3	3	4
Diagnosis *	D	2	2	1	0	3	3	4
	E	3	3	3	3	0	1	4
	F	3	3	3	3	1	0	4
	G	4	4	4	4	4	4	0

* Key to diagnostic groups:	
A anxiety neurosis	D tension headache
B depressive neurosis	E schizophrenia
C insomnia	F affective psychosis
	G alcohol or drug dependence

tests and be represented or described by a profile of test scores. In this situation the investigator is concerned with the proximity between profiles of test scores or symptom patterns.

The similarity of or distance between any two raters assessing several subjects can be calculated using a two-stage procedure. First one can define a measure of similarity or distance between any two profiles using the methods described in Sections 2.2, 2.3 or 2.4. The Euclidean distance between two profiles of k interval scores could, for example, be defined as follows:

$$D_{12}^{ii} = \left[\sum_{j=1}^{k} (x_{i1j} - x_{i2j})^2 \right]^{1/2} \qquad (2.10)$$

where x_{i1j} and x_{i2j} correspond to the jth score of subject i as produced by rater 1 and rater 2, respectively. The overall distance between raters 1 and 2 could then be obtained by summing or averaging the D_{12}^{ii}.

In the construction of a distance measure for the comparison of profiles the investigator might not wish to give equal prominence to each of the k test scores. Some form of weighted Euclidean distance might be appropriate:

$$D_{12}^{ii} = \left[\sum_{j=1}^{k} w_j(x_{i1j} - x_{i2j})^2 \right]^{1/2} \qquad (2.11)$$

In the more general case (that is, including categorical and ordinal responses) the investigator will need to define a series of disagreement weights for each possible pair of values obtained on each of the k measurements or assessments. The distance between rater 1 and rater 2 on the assessment of subject i can then be defined as the weighted sum of this subject's disagreement weights; that is,

$$D^{ii} = \sum_{j=1}^{k} w_j D_j^{ii} \qquad (2.12)$$

where w_j is the perceived importance (weight) of measurement or characteristic j (not dependent on which subject is being assessed) and D_j^{ii} is the disagreement weight (or distance between raters 1 and 2) for the jth measurement on subject i.

Consider a psychiatric assessment based on recording the presence (or absence) of k behavioural abnormalities or symptoms. If two raters agree on the presence or absence of a particular symptom then $D_j^{ii} = 0$, otherwise $D_j^{ii} = 1$. If all symptoms are considered to be equally important then one could let $w_j = 1$ for all of the k ratings. In this case the overall disagreement between raters is simply the sum of the disagreements for the separate symptoms. One might, however, think that disagreement over psychotic symptoms is more important than disagreement over symptoms such as headache or sleep problems and therefore change the w_j to convey this information. Instead of simply deciding whether a symptom is present the psychiatrists might give the patient a subjective rating for each symptom (ranging, for example, from 0 indicating absence to 5 indicating frequent or severe) and in this case each of the D_{12}^{ii} would have to be specified accordingly.

Returning to a profile of interval scores, another candidate for an agreement or similarity between two profiles is the Pearson product–moment correlation coefficient. If the profiles are very similar in shape the correlation will approach 1, if they are very different in shape they will approach 0 or even be negative. A disagreement index can be constructed by subtracting the correlation for +1. Despite the apparent simplicity of the correlation coefficient as an index of similarity, however, it has several limitations (see Section 2.9; and see also Section 3.4 of Dunn and Everitt, 1982).

2.6 Rater uncertainty

Suppose that clinicians are asked to place psychiatric patients into the diagnostic categories given in Table 2.4. This problem may be more complicated than has been described in Section 2.4 for at least two reasons. First, the categories may not be mutually exclusive. A psychiatrist may consider that a patient is suffering from both anxiety neurosis and alcoholism. This problem might be solved by considering the six diagnoses as a profile of six binary responses and dealing with the similarity between these profiles as suggested in Section 2.5. The second complication is rater uncertainty. The psychiatrist may think that a patient is suffering from either anxiety neurosis or depressive neurosis, but there may be difficulties in deciding which.

Consider the situation where there are possible categories A, B, ..., K. The problem of rater uncertainty can be overcome by asking the rater to rank order the appropriateness of these categories for the subject being assessed. A clear-cut decision to place a patient in category A without any uncertainty would be represented by giving the rank 1 to A and equal ranks of $\frac{1}{2}(k + 2)$ to the others. The value $\frac{1}{2}(k + 2)$ is simply the average of the ranks 2, 3, ..., k. If the rater thought that the appropriate category was either C or D, but was not sure which, then this could be represented by giving the ranks 1.5 to each of C and D and $\frac{1}{2}(k + 3)$ to the other $k - 2$ response categories. If, however, the rater could not decide between C and D, but thought D to be the most likely, then this could be represented by giving D rank 1, C rank 2 and the rest rank $\frac{1}{2}(k + 3)$. Kraemer (1980) has suggested that agreement between two profiles of ranks produced in this way should be measured by the Spearman rank correlation coefficient between them. In the relatively simple situations where only two categories might be confused (say, A and B) there are five possible profiles of ranks. A unique choice of A, a unique choice of B, an equivocal response A/B (meaning A or B, with equal rank), an ordered response AB (meaning A with rank 1 and B with rank 2), and the ordered response BA. Kraemer (1980) has produced a table of the degree of agreement beween these responses as a function of the number of categories, k. This is shown in Table 2.5. Note that perfect agreement, with a value of 1.0, is indicated when two profiles of ranks are identical in every respect. Note also that it is possible to obtain negative values for this agreement index. A disagreement index could be constructed by subtracting the corresponding value of Spearman's rho from $+1$.

Table 2.5 Degree of agreement between unique (A), equivocal (A/B), and ranked (AB, BA) observations as a function of the number of response categories (k)

		k				
	Spearman rho	2	3	5	10	∞
A vs A/B	$\{(k - 2)/2(k - 1)\}^{1/2}$	0.00	0.50	0.61	0.67	0.71
A vs AB	$\{k/(k - 1)\}^{1/2}$	1.00	0.87	0.79	0.74	0.71
A vs BA	$\{k^2(k - 3)/2(k - 1)^3\}^{1/2}$	-1.00	0.00	0.40	0.58	0.71
A/B vs AB or BA	$\{k(k - 2)/(k - 1)^2\}^{1/2}$	0.00	0.87	0.97	0.99	1.00
AB vs BA	$(k^2 - 2k - 1)/(k - 1)^2$	—	0.50	0.88	0.98	1.00

2.7   Chance agreement

The idea of chance agreement to be discussed in this chapter is not based on the behaviour of populations of measurements, raters or subjects but on random permutations of the sample of measurements or ratings *actually observed*. Consider the weights given in Table 1.1 or the examination marks in Table 2.2. Measures of agreement or disagreement for the data in these tables are based on a knowledge of the matching pairs of observations (because they are known to be made on the same objects or individuals). But what would happen if it were not known which pairs of measurements correspond to the individuals being measured and the measurements from the first instrument or rater were in fact randomly matched to those of the second instrument or rater? This would provide a measure of agreement that has arisen purely by chance. If this process could be repeated over and over again, and the agreement calculated for each random matching of measurements, a probability distribution of agreement indices could be generated. One could also obtain an average or expected value for chance agreement (see Appendix 1).

To illustrate this idea the data in Table 2.2 will be presented in a different way. Consider Table 2.6. This is a 29×29 distance matrix recording the absolute distances between Examiner A's ratings on the 29 candidates with each of the 29 ratings of Examiner B. The diagonal entries correspond to the absolute values of the differences recorded in the last column of Table 2.2. These are the distances between the Examiners' ratings on the *same* candidates. The off-diagonal entries correspond to the distances between Examiners when *different* candidates are being compared. Entry (4, 5) for example is the distance between Examiner B's rating of candidate 4 and Examiner A's rating of candidate 5 (that is $|2 - 0| = 2$). Entry $(29, 1)$ is the distance between Examiner B's rating of candidate 29 and Examiner A's rating of the 1st candidate (that is, $|3 - 1| = 2$). If the distance matrix given in Table 2.6 is referred to as the matrix $\mathbf{X}$, then the observed disagreement between Examiners A and B is the trace of $\mathbf{X}$ (that is $\sum_{i=1}^{29} X_{ii}$, where X_{ij} is the i,jth entry of the matrix $\mathbf{X}$).

Random matching of the ratings of Examiner A with those of Examiner B can be achieved by randomly permuting the columns (or the rows) of the distance matrix $\mathbf{X}$. For each random permutation the trace of the new distance matrix can be determined. This would represent chance disagreement. If the random permutation process were repeated over and over again this would yield a probability distribution for chance disagreement. Alternatively, the level of disagreement could be calculated for all possible permutations of the columns of Table 2.6 (but there are 29! possible permutations, so this is a rather impractical exercise). Table 2.7 lists the levels of disagreement (in the order obtained) for 99 random permutations of the columns in Table 2.6.

Table 2.6 Matrix of distances (disagreement weights) for 29 examinees ratings shown in Table 2.2*

												Examiner A																
1	2	2	0	2	2	2	2	2	0	1	0	2	2	2	1	2	1	0	2	0	2	2	2	2	2	1	1	0
1	0	0	2	0	4	0	0	0	2	1	2	0	4	4	1	0	1	2	0	2	4	0	0	4	0	1	3	2
1	0	0	2	0	4	0	0	0	2	1	2	0	4	4	1	0	1	2	0	2	4	0	0	4	0	1	3	2
1	2	2	0	2	2	2	2	2	0	1	0	2	2	2	1	2	1	0	2	0	2	2	2	2	2	1	1	0
1	0	0	2	0	4	0	0	0	2	1	2	0	4	4	1	0	1	2	0	2	4	0	0	4	0	1	3	4
2	3	3	1	3	1	3	3	3	1	2	1	3	1	1	2	3	2	1	3	1	1	3	3	1	3	2	0	1
1	0	0	2	0	4	0	0	0	2	1	2	0	4	4	1	0	1	2	0	2	4	0	0	4	0	1	3	2
1	0	0	2	0	4	0	0	0	2	1	2	0	4	4	1	0	1	2	0	2	4	0	0	4	0	1	3	2
1	0	0	2	0	4	0	0	0	2	1	2	0	4	4	1	0	1	2	0	2	4	0	0	4	0	1	3	2
2	3	3	1	3	1	3	3	3	1	2	1	3	1	1	2	3	2	1	3	1	1	3	3	1	3	2	0	1
1	2	2	0	2	2	2	2	2	0	1	0	2	2	2	1	2	1	0	2	0	2	2	2	2	2	1	1	0
2	3	3	1	3	1	3	3	3	1	2	1	3	1	1	2	3	2	1	3	1	1	3	3	1	3	2	0	1
0	1	1	1	1	3	1	1	1	1	0	1	1	3	3	0	1	0	1	1	1	3	1	1	3	1	0	2	1
2	3	3	1	3	1	3	3	3	1	2	1	3	1	1	2	3	2	1	3	1	1	3	3	1	3	2	0	1
2	3	3	1	3	1	3	3	3	1	2	1	3	1	1	2	3	2	1	3	1	1	3	3	1	3	2	0	1
1	2	2	0	2	2	2	2	2	0	1	0	2	2	2	1	2	1	0	2	0	2	2	2	2	2	1	1	0
1	2	2	0	2	2	2	2	2	0	1	0	2	2	2	1	2	1	0	2	0	2	2	2	2	2	1	1	0
1	2	2	0	2	2	2	2	2	0	1	0	2	2	2	1	2	1	0	2	0	2	2	2	2	2	1	1	0
2	3	3	1	3	1	3	3	3	1	2	1	3	1	1	2	3	2	1	3	1	1	3	3	1	3	2	0	1
1	0	0	2	0	4	0	0	0	2	1	2	0	4	4	1	0	1	2	0	2	4	0	0	4	0	1	3	2
2	3	3	1	3	1	3	3	3	1	2	1	3	1	1	2	3	2	1	3	1	1	3	3	1	3	2	0	1
3	4	4	2	4	0	4	4	4	2	3	2	4	0	0	3	4	3	2	4	2	0	4	4	0	4	3	1	2
1	0	0	2	0	4	0	0	0	2	1	2	0	4	4	1	0	1	2	0	2	4	0	0	4	0	1	3	2
1	0	0	2	0	4	0	0	0	2	1	2	0	4	4	1	0	1	2	0	2	4	0	0	4	0	1	3	2
2	3	3	1	3	1	3	3	3	1	2	1	3	1	1	2	3	2	1	3	1	1	3	3	1	3	2	0	1
1	2	2	0	2	2	2	2	2	0	1	0	2	2	2	1	2	1	0	2	0	2	2	2	2	2	1	1	0
1	2	2	0	2	2	2	2	2	0	1	0	2	2	2	1	2	1	0	2	0	2	2	2	2	2	1	1	0
3	4	4	2	4	0	4	4	4	2	3	2	4	0	0	3	4	3	2	4	2	0	4	4	0	4	3	1	2
2	3	3	1	3	1	3	3	3	1	2	1	3	1	1	2	3	2	1	3	1	1	3	3	1	3	2	0	1

(The rows correspond to Examiner B.)

* Columns represent the candidates 1–29 and the rows correspond to the same 29 candidates taken in the same order.

Table 2.7 Disagreement indices * for 99 random permutations of the columns of Table 2.6

56	56	46	58	46	44	42	46	58
50	48	42	44	46	52	64	50	50
40	46	50	50	50	52	48	54	46
52	56	60	44	42	42	54	48	42
52	44	34	44	44	50	54	54	58
44	48	50	54	44	46	40	38	48
44	42	40	46	38	42	40	54	44
46	38	36	52	52	50	48	52	42
52	50	44	50	58	38	40	36	48
56	46	52	44	42	52	50	44	48
52	54	52	52	64	50	46	38	58

Minimum value:	34
Maximum value:	64
Mean:	47.9
Median:	48
Standard deviation:	6.23

* Using linear disagreement weights

The mean value for this sample of 99 values is 47.9. Although they have not been enumerated, the mean value of the 29! possible permutations can be obtained for the mean of all of the entries in Table 2.6. This mean is 47.7.

If the observed disagreement (20) is compared with the 99 chance agreements listed in Table 2.7, it can be seen that it is considerably smaller than the minimum disagreement observed from these random permutations (34). These results suggest that the probability of observing a disagreement of 20 by chance is considerably less than $\frac{1}{100}$. An exact probability could be obtained by complete enumeration of the 29! possible disagreements, and a more exact approximation could be obtained through the use of considerably more than the 99 permutations used to generate the disagreements given in Table 2.7.

The above procedure defines a statistical significance test called a *permutation* or *randomization test*. On the assumption that it is possible to define a measure of distance or similarity between all pairs of ratings (whether nominal, ordinal or interval measures) it is possible to use this random permutation procedure to generate a significance level for a particular level of disagreement or agreement between two raters. This type of test can be quite useful when the form of the probability distribution of a particular agreement or disagreement measure is not known, or simply to illustrate what is meant (in this particular situation, at least) by chance agreement.

2.8 Chance-corrected agreement

In general, a chance-corrected index of agreement can be defined as follows:

$$\text{agreement} = 1 - \frac{\text{observed disagreement}}{\text{disagreement expected by chance}} \qquad (2.13)$$

This will have a value of zero when the observed disagreement is merely chance, unity when there is perfect agreement, and will be negative if disagreement is worse than chance. For the example discussed in the last section (the data in Tables 2.2 and 2.7) the chance-corrected measure of agreement between the two examiners is

$$\text{agreement} = 1 - \frac{20}{47.7}$$

$$= 0.58$$

Now reconsider Table 2.3. Using the linear disagreement weights in Table 2.3(b) the observed measure of disagreement is 20 (q.v.). If examiners are assumed to be independent one can work out expected values for the counts in Table 2.3(a) from the marginal probabilities. These expected counts are given in Table 2.8. The weighted sum of these expected counts (using the linear weights in Table 2.3(b)) is 47.6 which, apart from a trivial rounding error, is the same as that obtained from the average distance Table 2.6. The agreement index defined in Equation (2.13) is equivalent to Cohen's *weighted kappa statistic* (Cohen, 1968). Using the squared differences (quadratic disagreement weights) in Table 2.3(b) the weighted κ statistic for Table 2.3(a) is $(1 - 24/120.56) = 0.80$.

Table 2.8 Table of chance-expected counts derived from Table 2.3(a) on the assumption of independence of Examiners A and B

		Rating by Examiner B					
		0	1	2	3	4	Total
	0	3.724	0.414	3.310	3.724	0.828	12
Rating by	1	1.552	0.172	1.379	1.552	0.345	5
Examiner A	2	1.862	0.207	1.655	1.862	0.414	6
	3	0.310	0.035	0.276	0.310	0.069	1
	4	1.552	0.172	1.379	1.552	0.345	5
Total		9	1	8	9	2	29

Now consider a more general square contingency table of counts (rating by first observer cross-classified by rating by second observer). This is shown in Table 2.9. The total number of subjects or objects being classified is n. Part (a) gives the observed number of subjects (n_{ij}) that are assigned to category i by the first observer and category j by the second. Part (b) gives the corresponding counts that are expected by chance and, finally, part (c) gives the corresponding disagreement weights.

Following the definition given in Equation (2.13) the chance-corrected measure of agreement is given by

$$\text{agreement} = 1 - \frac{\sum_i \sum_j n_{ij} d_{ij}}{\sum_i \sum_j e_{ij} d_{ij}} \qquad (2.14)$$

When the e_{ij} are given by

$$e_{ij} = \frac{1}{n} n_{i+} n_{+j} \qquad (2.15)$$

Table 2.9 Square contingency table for two raters

(a) Observed counts

		Rater 2 category				
		1	2	j	c	Row (rater 1) totals
Rater 1 category	1	n_{11}	n_{12}	n_{1j}	n_{1c}	n_{1+}
	2	n_{21}	n_{22}	n_{2j}	n_{2c}	n_{2+}
	i	n_{i1}	n_{i2}	n_{ij}	n_{ic}	n_{i+}
	c	n_{c1}	n_{c2}	n_{cj}	n_{cc}	n_{c+}
Column (rater 2) totals		n_{+1}	n_{+2}	n_{+j}	n_{+c}	n

(b) Expected counts

		Rater 2 category			
		1	2	j	c
Rater 1 category	1	e_{11}	e_{12}	e_{1j}	e_{1c}
	2	e_{21}	e_{22}	e_{2j}	e_{2c}
	i	e_{i1}	e_{i2}	e_{ij}	e_{ic}
	c	e_{c1}	e_{c2}	e_{cj}	e_{cc}

(c) Disagreement weights $(d_{ij} = d_{ji})$

		Rater 2 category			
		1	2	j	c
Rater 1 category	1	d_{11}	d_{12}	d_{1j}	d_{1c}
	2	d_{21}	d_{22}	d_{2j}	d_{2c}
	i	d_{i1}	d_{i2}	d_{ij}	d_{ic}
	c	d_{c1}	d_{c2}	d_{cj}	d_{cc}

the agreement index is Cohen's weighted κ coefficient (q.v.). If the $d_{ij} = 0$ when $i = j$ and $d_{ij} = 1$ when $i \neq j$ then the agreement index is Cohen's unweighted κ coefficient (Cohen, 1960).

If

$$d_{ij} = (i - j)^2 \qquad (2.16)$$

then quadratic disagreement weights (q.v.) are being used. Schouten (1985, 1986) has suggested that this particular weighted κ statistic be called v.

If, in an analogous way to Equations (2.1) and (2.2), we define an agreement weight, s_{ij}, by

$$s_{ij} = 1 - \frac{d_{ij}}{\max(d_{ij})} \qquad (2.17)$$

then

$$0 \leq s_{ij} = s_{ji} \leq 1 = s_{ii} \qquad (2.18)$$

and

$$\text{observed agreement} = \sum_i \sum_j n_{ij} s_{ij} \qquad (2.19)$$

Similarly

$$\text{expected agreement} = \sum_i \sum_j e_{ij} s_{ij} \tag{2.20}$$

and

$$\frac{\text{observed agreement} - \text{expected agreement}}{n - \text{expected agreement}} = \frac{\sum_i \sum_j n_{ij} s_{ij} - \sum_i \sum_j e_{ij} s_{ij}}{n - \sum_i \sum_j e_{ij} s_{ij}}$$

$$= \frac{\sum_i \sum_j n_{ij}\left(1 - \dfrac{d_{ij}}{\max(d_{ij})}\right) - \sum_i \sum_j e_{ij}\left(1 - \dfrac{d_{ij}}{\max(d_{ij})}\right)}{n - \sum_i \sum_j \left(1 - \dfrac{d_{ij}}{\max(d_{ij})}\right)}$$

$$= \frac{n - \dfrac{1}{\max(d_{ij})}\begin{bmatrix} \text{observed} \\ \text{disagreement} \end{bmatrix} - n + \dfrac{1}{\max(d_{ij})}\begin{bmatrix} \text{expected} \\ \text{disagreement} \end{bmatrix}}{n - n + \dfrac{1}{\max(d_{ij})}\begin{bmatrix} \text{expected} \\ \text{disagreement} \end{bmatrix}}$$

$$= 1 - \frac{\text{observed disagreement}}{\text{expected disagreement}} \tag{2.21}$$

If observed and expected agreement are represented by proportions rather than counts then Equation (2.21) is equivalent to the more familiar definition of weighted κ. That is

$$\text{weighted } \kappa = \frac{\sum_i \sum_j (n_{ij}/n) s_{ij} - \sum_i \sum_j (e_{ij}/n) s_{ij}}{1 - \sum_i \sum_j (e_{ij}/n) s_{ij}} \tag{2.22}$$

When $s_{ij} = 0$ if $i \neq j$ and $s_{ij} = 1$ if $i = j$ then Equation (2.22) reduces to the familiar definition of the unweighted κ statistic.

The mean of the ratings over the n subjects for the first observer is given by $\bar{x}_1 = \sum_i i n_{i+}/n$ and the mean of the n ratings for the second observer by $\bar{x}_2 = \sum_i i n_{+i}/n$.

Similarly, the corresponding sums of squares are $s_1^2 = \sum_i n_{i+}(i - \bar{x}_1)^2$ and $s_2^2 = \sum_i n_{+i}(i - \bar{x}_2)^2$. Finally, the sum of cross-products for the two raters is given by $s_{12} = \sum_i \sum_j n_{ij}(i - \bar{x}_1)(j - \bar{x}_2)$.

An alternative expression for this particular weighted κ,

$$\text{agreement} = 1 - \frac{\sum_i \sum_j n_{ij}(i - j)^2}{(1/n) \sum_i \sum_j n_{i+} n_{+j}(i - j)^2} \tag{2.23}$$

is given by

$$\text{agreement} = \frac{2 s_{12}}{s_1^2 + s_2^2 + n(\bar{x}_1 - \bar{x}_2)^2} \tag{2.24}$$

Equation (2.24) was first devised by Krippendorff (1970), but can also be found in Schouten (1985, 1986). This formula can, of course, be used to provide a measure of agreement between two sets of continuous measurements such as the weights given in Table 1.1. It is very closely related to the intra-class correlation coefficient discussed in the following section.

2.9   Intra-class correlation

In this section the concept of an intra-class correlation will be introduced, with particular attention being paid to the agreement between two instruments, observers or raters. For the case of several measuring instruments there are several different forms of the intra-class correlation. These will be discussed in detail in Chapter 6. First, consider a simple way of actually calculating an intra-class correlation coefficient. Pearson's (1901) intra-class correlation between N pairs of observations X_{i1} and X_{i2} is equivalent to the Pearson product-moment correlation between $2N$ pairs of observations, the first N of which are the original observations, and the second N the same observations with X_{i1} replacing X_{i2} and vice versa. If, for example, the first three packets from Table 1.1 are considered, the intra-class correlation is calculated from the Pearson product-moment correlation of the following six pairs of measurements:

560	550	550	560
140	140	140	140
1680	1710	1710	1680

Its value is 0.9996. The corresponding intra-class correlation derived from the data on examination ratings in Table 2.2 is 0.80. Note that this is the same as the weighted κ statistic obtained through the use of quadratic disagreement weights.

In the case of two measuring instruments or raters the value of the intra-class correlation depends in part on the corresponding product-moment correlation, but it is also dependent on the differences between the means and the standard deviations of the two variables. If the product-moment correlation is represented by r and the Pearson intra-class correlation by r_i, then

$$r_i = \frac{\{\sum (s_1^2 + s_2^2) - (s_1 - s_2)^2\}r - (\bar{x}_1 - \bar{x}_2)^2/2}{(s_1^2 + s_2^2) + (\bar{x}_1 - \bar{x}_2)^2/2} \tag{2.25}$$

where $\bar{x}_1$ and $\bar{x}_2$ are the mean values for instrument 1 and instrument 2, respectively, and s_1 and s_2 are the corresponding standard deviations. The two correlations r_i and r will only be the same when $\bar{x}_1 = \bar{x}_2$ *and* $s_1 = s_2$, otherwise $r_i < r$. The intra-class correlation, r_i, can be defined by the following algebraic expression:

$$r_i = \frac{2 \sum_{i=1}^{N}(x_{i1} - m)(x_{i2} - m)}{\sum_{i=1}^{N}(x_{i1} - m)^2 + \sum_{i=1}^{N}(x_{i2} - m)^2} \tag{2.26}$$

where $m = (\bar{x}_1 + \bar{x}_2)/2$. The equivalent form of Equation (2.24) is

$$\text{agreement index} = \frac{2 \sum_{i=1}^{N}(x_{i1} - \bar{x}_1)(x_{i2} - \bar{x}_2)}{\sum_{i=1}^{N}(x_{i1} - \bar{x}_1)^2 + \sum_{i=1}^{N}(x_{i2} - \bar{x}_2)^2 + \sum_{i=1}^{N}(\bar{x}_1 - \bar{x}_2)^2} \tag{2.27}$$

Expressions (2.26) and (2.27) are not identical, but for the data in Table 2.2 they do, in fact, give identical results when expressed to only two decimal places. Fleiss and Cohen (1973) have demonstrated that Cohen's weighted κ (using quadratic weights) is approximately equivalent to an intra-class correlation.

Krippendorff (1970) has demonstrated that the intra-class correlation as defined by Equation (2.27) is equivalent to a weighted κ-like statistic defined as in Equations (2.18) and (2.28) but where

$$e_{ij} = \frac{1}{n}\left(\frac{n_{i+} + n_{+i}}{2}\right)\left(\frac{n_{j+} + n_{+j}}{2}\right) \qquad (2.28)$$

Here the expected frequencies are calculated on a joint marginal distribution. That is, there is an assumption of marginal homogeneity for the two examiners. If Equation (2.28) applies and the disagreement weights are chosen such that $d_{ij} = 0$ for $i = j$ and $d_{ij} = 1$ if $i \neq j$, then the resulting coefficient defined by (2.14) is equivalent to Scott's index of inter-coder agreement, π (Scott, 1955). The reader is referred to Chapter 7 for further discussion of intra-class correlations for categorical data.

2.10   Monotonic measures of agreement for ranked data

Suppose two judges (A and B) are asked to independently rank four student essays. The four essays are represented by the symbols a, b, c, and d. The ranks given to these essays by the two judges are these:

Essay	a	b	c	d
Judge A	4	1	3	2
Judge B	3	1	2	4

How close are these two profiles of ranks? In Section 2.6 it was suggested that the similarity of two sets of ranks could be estimated using Spearman's rank correlation coefficient (ρ). Spearman's ρ and also weighted κ coefficients using linear or quadratic disagreement weights are examples of rank-interval statistics. That is, they involve assumptions concerning the size of the intervals between pairs of ranks, or between pairs of ordinal categories. Spearman's ρ for example, treats 'rank' as if it is a unit of measurement so that the interval between ranks 7 and 9 is taken to be 2 units, the interval between ranks 1 and 5 is 4 units, and so on. Weighted κ statistics involve similar assumptions concerning the sizes of the disagreement weights.

An alternative to Spearman's ρ is Kendall's (1938) rank correlation coefficient (τ). Kendall's τ does not involve any assumptions concerning the exact size of the intervals between pairs of ranks. Another alternative is Goodman and Kruskal's (1954) γ. These two statistics measure *monotonic* agreement. If essay a is higher than essay b according to judge A and is also higher than essay b according to Judge B, then there is agreement between the two judges on the relative ranks of essays a and b. Disagreement between the two judges occurs when one of them ranks essay a above essay b and the other ranks b above a.

To calculate τ we first obtain the number of pairs of essays which are ranked in the same order by judges A and B. (Let this equal P.) Similarly, let Q be the number of essays in which the rankings are in the opposite order. Then τ is defined by

$$\tau = \frac{P - Q}{\frac{1}{2}n(n - 1)} \qquad (2.29)$$

where n is the total number of essays marked. For the above example $\tau = 0.33$. When there are tied observations they are given the average of the ranks they would have received if there were no ties. The effect of ties is to modify the formula for τ. In this case τ is defined by

$$\tau = \frac{P - Q}{\sqrt{(\tfrac{1}{2}n(n-1) - T_A)}\,\sqrt{(\tfrac{1}{2}n(n-1) - T_B)}} \tag{2.30}$$

where $T_A = \tfrac{1}{2}\sum_i t_i(t_i - 1)$, t_i being the number of tied observations on the ith group of ties of Judge A; and $T_B = \tfrac{1}{2}\sum_j t_j(t_j - 1)$, t_j being the number of tied observations in the jth group of ties of Judge B. For an example of the use of (2.30) for data with tied observations, reconsider Table 2.3(a). Here

$$T_A = 101, \; T_B = 101 \quad \text{and} \quad \tfrac{1}{2}n(n-1) = 406$$

$$P = 263 \quad \text{and} \quad Q = 4$$

Therefore

$$\tau = \frac{259}{406 - 101} = 0.85$$

Stavig (1984) has suggested the use of the monotonic equivalent of the intra-class correlation coefficient as a measure of total agreement between raters. He points out that the simple τ-statistic suffers for the same defects as the Pearson product-moment correlation coefficient as a measure of agreement. Stavig's transpositional measure of monotonic agreement is obtained by including each subject twice in the calculation of τ in a similar way to that used in the calculation of Pearson's intra-class correlation coefficient (see Section 2.9). This will be illustrated by reference to the data in Table 2.3(a). The transposed table is given in Table 2.10(a). The sum of the entries of

Table 2.10

(a) Transposed equivalent of Table 2.3(a)

		Examiner A				
		0	1	2	3	4
	0	9	0	0	0	0
	1	1	0	0	0	0
Examiner B	2	2	5	1	0	0
	3	0	0	5	0	4
	4	0	0	0	1	1

(b) Sum of counts in Table 2.3(a) and 2.10(b)

		1st rating					
		0	1	2	3	4	Total
	0	18	1	2	0	0	21
	1	1	0	5	0	0	6
2nd rating	2	2	5	2	5	0	14
	3	0	0	5	0	5	10
	4	0	0	0	5	2	7
Total		21	6	14	10	7	58

Table 2.3(a) and Table 2.10(a) is given in Table 2.10(b). Stavig refers to this type of table as the transposition of the original data. Note the symmetry of Table 2.10(b). The τ statistic is now calculated to produce the rank-monotonic intra-class correlation of agreement.

Here

$$T_A = T_B = 382$$

$$\tfrac{1}{2}n(n - 1) = 1653$$

$$P = 1056 \qquad Q = 79$$

$$\tau_i = \frac{1056 - 79}{1653 - 382}$$

$$= 0.77$$

A weighted κ statistic could also be calculated from Table 2.10(b). This will be left as an exercise for the reader.

2.11 What is good agreement?

Landis and Koch (1977a) have given some arbitrary 'benchmarks' for the evaluation of observed κ values. These are as follows:

κ	*strength of agreement*
0.00	poor
0.01–0.20	slight
0.21–0.40	fair
0.41–0.60	moderate
0.61–0.80	substantial
0.81–1.00	almost perfect

In the view of the present author these are too generous, but any series of standards such as these are bound to be subjective. Given that the choice of weights for a κ statistic is also a subjective decision, then it should be clear that there is no simple answer to the question: 'How good is agreement?'

For the data in Table 2.2 at least four coefficients can be generated. A simple unweighted κ using the raw data is 0.21. The κ coefficient obtained from Table 2.1 (agreement in pass or fail) is 0.54. The two weighted κs from Table 2.2 are 0.58 and 0.80 (using linear and quadratic weights, respectively). Which κ is the most useful? The examiners might prefer the last one as this almost brings them into the 'almost perfect' range. The candidates on the other hand would be concerned by the 'moderate' expressed by the second index (0.54). Clearly the interpretation of the results is dependent upon the uses to which the measurements are to be put. If a candidate is to be finally assessed using the average marks of the two examiners in the above example, then perhaps agreement is good enough. If, in future, candidates are going to be assessed by either one of the examiners, but not both, then perhaps the agreement is nowhere near good enough. In the latter case sources of disagreement should be searched for and rectified. Only then might one have sufficient confidence in the assessment of a lone examiner.

One important factor in the assessment of the size of a κ coefficient (or any other index of agreement) is the maximum value it could have had for a given set of data. For the data in Table 2.3(a), for example, it is sensible to ask 'What is the maximum possible level of agreement given the marginal totals for each examiner?' Consider each diagonal cell separately. In cell $(0, 0)$ the maximum number of candidates that it could contain given the marginal totals is 9. For cell $(1, 1)$ the corresponding count is 1. For $(2, 2)$, $(3, 3)$ and $(4, 4)$ the counts are 6, 1 and 2, respectively. The maximum for the sum of the counts in these diagonal cells is 19. That is, if agreement = 1 for diagonal entries and 0 otherwise, then

$$\text{maximum agreement} = \sum_i \min(n_{i+}, n_{+i}) \qquad (2.31)$$

This will only reach the theoretical maximum when the margins are homogeneous (that is, when $n_{i+} = n_{+i}$ for all i). In the case of the simple (unweighted) κ, Table 2.3(a) yields

$$\max(\kappa) = 1 - \frac{\text{minimum disagreement}}{\text{expected disagreement}} \qquad (2.32)$$

$$= \frac{\text{maximum agreement} - \text{expected agreement}}{n - \text{expected agreement}} \qquad (2.33)$$

$$= \frac{19 - 6.21}{29 - 6.21}$$

$$= 0.56$$

$$\frac{\kappa}{\max(\kappa)} = \frac{0.21}{0.56} = 0.38$$

Note that

$$\frac{\kappa}{\max(\kappa)} = \frac{\text{observed agreement} - \text{expected agreement}}{\text{maximum agreement} - \text{expected agreement}} \qquad (2.34)$$

This is a general measurement of agreement as defined, for example, by Brennan and Prediger (1981) and Hubert and Golledge (1983).

2.12   Exercises

2.1   For the assessment of agreement between binary responses, consider the two matching coefficients S'_{AB}, and S_{AB}, defined by Equations (2.7) and (2.8), respectively. Demonstrate that, if one allows for chance agreement for each of these coefficients by the following,

$$\text{chance-corrected coefficient} = \frac{\text{observed coefficient} - \text{expected coefficient}}{1 - \text{expected coefficient}}$$

then the result is identical to Cohen's κ statistic in both cases (see Fleiss, 1975).

2.2   Consider the following table of agreement for binary responses

Rater B

		Yes	No	Total
	Yes	15	5	20
Rater A	No	5	35	40
	Total	20	40	

Show that Cohen's κ for this data is identical to the intra-class correlation and to the product-moment correlation coefficient, φ (see Bartko and Carpenter, 1976).

2.3   In 1940, Lundberg (1940) presented the results of an experiment to answer the question 'What is the degree of agreement in "commonsense" judgements of socio-economic status of the population of a community by two persons who are themselves of radically different status, that is, how are informal ratings of socio-economic status influenced by the socio-economic status of the rater?' Some of Lundberg's results were discussed by Robinson (1957) who presented the information in Table 2.11.

Table 2.11 Relation between janitor's and banker's ratings of the socioeconomic status of 196 families

		Janitor's rating					
		1	2	3	4	5	6
	6	0	0	0	0	0	0
	5	0	0	0	6	8	8
	4	0	1	0	21	27	0
Banker's rating	3	0	1	25	47	13	2
	2	0	4	6	4	1	0
	1	3	4	11	3	1	0

Investigate the levels of agreement between these two raters using weighted κ coefficients, intra-class correlation and rank-based correlation coefficients.

2.4   Calculate Pearson's product-moment and intra-class correlation coefficients for the two columns of measurements in Table 1.1. Compare these results with Kendall's τ and its corresponding intra-class coefficient.

2.5   Investigate levels of agreement between pairs of raters for the estimates of string length given in Table 1.2.

3

Reliability, Consistency and Stability

3.1 Introduction to reliability

The classical test theory model for psychological measurements is that given in Equation (1.1), that is

$$X_{ij} = \tau_i + e_{ij} \tag{3.1}$$

where X_{ij} represents the jth measurement on the ith subject. In the context of psychometric theory the measurements are usually assumed to be scores obtained from a psychological test such as one of the many tests of cognitive ability. τ_i is defined to be the *true* score for subject i and e_{ij} the corresponding error of measurement. If the observed score X_{ij} is regarded as one example of the infinitely large number of continuous scores that might have been obtained at the jth assessment of subject i (that is, the *universe* of possible test scores for subject i), then the subject's true score, τ_i, is defined as the expected value of X_{ij}

$$\underset{j}{E}(X_{ij}) = \tau_i \tag{3.2}$$

Note that, as in Chapter 1, true scores are defined as expected values of the corresponding observed scores. If one were to take expectations over the universe of possible measurements *and* population of subjects, then

$$\underset{i,j}{E}(X_{ij}) = \underset{i}{E}(\tau_i) = \mu \tag{3.3}$$

where μ is the mean of all possible scores, X_{ij}. It follows that

$$\underset{i,j}{E}(e_{ij}) = \underset{j}{E}(e_{ij}) = 0 \tag{3.4}$$

It is assumed that measurement errors are uncorrelated with each other and with the true scores (both within and across subjects). It follows that

$$\text{Var}(X_{ij}) = \text{Var}(\tau_i) + \text{Var}(e_{ij}) \tag{3.5}$$

and

$$\begin{aligned}
\text{Cov}(X_{ij}, \tau_i) &= \underset{i,j}{E}(X_{ij}\tau_i) - \underset{i,j}{E}(X_{ij})\underset{i}{E}(\tau_i) \\
&= \underset{i,j}{E}[(\tau_i + e_{ij})\tau_i] - \underset{i,j}{E}(\tau_i + e_{ij})\underset{i}{E}(\tau_i) \\
&= \underset{i}{E}(\tau_i^2) + \underset{i,j}{E}(e_{ij}\tau_i) - [\underset{i}{E}(\tau_i)]^2 - \underset{i,j}{E}(e_{ij})\underset{i}{E}(\tau_i) \\
&= \underset{i}{E}(\tau_i^2) - [\underset{i}{E}(\tau_i)]^2 \\
&= \text{Var}(\tau_i) \quad\quad\quad\quad\quad\quad\quad\quad\quad\quad (3.6)
\end{aligned}$$

The *reliability* of a test or measuring instrument is defined as the ratio of the true score variance to the observed score variance, that is

$$R = \frac{\text{Var}(\tau_i)}{\text{Var}(X_{ij})} \quad\quad\quad\quad (3.7)$$

$$= \frac{\text{Cov}(X_{ij}, \tau_i)}{\text{Var}(X_{ij})} \quad\quad\quad\quad (3.8)$$

This is also equivalent to

$$R = 1 - \frac{\text{Var}(e_{ij})}{\text{Var}(X_{ij})} \quad\quad\quad\quad (3.9)$$

Note the similarity of form of (3.9) with the agreement measures defined by Equation (2.13).

Now consider the correlation between a subject's observed score (X_{ij}) and the corresponding true score (τ_i).

$$\begin{aligned}
\text{Corr}(X_{ij}, \tau_i) &= \frac{\text{Cov}(X_{ij}, \tau_i)}{\sqrt{(\text{Var}(X_{ij}))}\sqrt{(\text{Var}(\tau_i))}} \\
&= \frac{\text{Var}(\tau_i)}{\sqrt{(\text{Var}(X_{ij}))}\sqrt{(\text{Var}(\tau_i))}} \\
&= \sqrt{\frac{\text{Var}(\tau_i)}{\text{Var}(X_{ij})}} = \sqrt{R} \quad\quad\quad (3.10)
\end{aligned}$$

If one has two parallel measurements for each subject i, X_{i1} and X_{i2}, then

$$X_{i1} = \tau_i + e_{i1} \quad\quad \text{and} \quad\quad X_{i2} = \tau_i + e_{i2} \quad\quad (3.11)$$

where $\text{Var}(X_{i1}) = \text{Var}(X_{i2}) = \text{Var}(X_{ij})$ and $\text{Var}(e_{i1}) = \text{Var}(e_{i2}) = \text{Var}(e_{ij})$. The correlation between X_{i1} and X_{i2} is given by

$$\begin{aligned}
\text{Corr}(X_{i1}, X_{i2}) &= \frac{\text{Cov}(X_{i1}, X_{i2})}{\sqrt{(\text{Var}(X_{i1}))}\sqrt{(\text{Var}(X_{i2}))}} \\
&= \frac{\text{Cov}(\tau_i + e_{i1}, \tau_i + e_{i2})}{\text{Var}(X_{ij})} \\
&= \frac{\text{Var}(\tau_i)}{\text{Var}(X_{ij})} \quad\quad\quad\quad (3.12) \\
&= R \quad\quad\quad\quad\quad\quad\quad (3.13)
\end{aligned}$$

Consider the sum of these two test scores

$$X_i = X_{i1} + X_{i2} \tag{3.14}$$

$$= 2\tau_i + e_{i1} + e_{i2} \tag{3.15}$$

$$\underset{i}{\text{E}}(X_i) = 2\tau_i$$

$$\text{Var}(X_i) = 4\,\text{Var}(\tau_i) + \text{Var}(e_{i1}) + \text{Var}(e_{i2})$$

$$= 4\,\text{Var}(\tau_i) + 2\,\text{Var}(e_{ij}) \tag{3.16}$$

But,

$$\text{Var}(2\tau_i) = 4\,\text{Var}(\tau_i) \tag{3.17}$$

and so the reliability of the sum, X_i, is given by

$$R' = \frac{4\,\text{Var}(\tau_i)}{4\,\text{Var}(\tau_i) + 2\,\text{Var}(e_{ij})}$$

$$= \frac{2\,\text{Var}(\tau_i)}{2\,\text{Var}(\tau_i) + \text{Var}(e_{ij})} \tag{3.18}$$

If both the top and bottom of this equation are now divided by $\text{Var}(X_{ij})$, this yields

$$R' = \frac{2\,\text{Var}(\tau_i)/\text{Var}(X_{ij})}{2\,\text{Var}(\tau_i)/\text{Var}(X_{ij}) + \text{Var}(e_{ij})/\text{Var}(X_{ij})}$$

$$= \frac{2R}{2R + (1 - R)}$$

$$= \frac{2R}{1 + R} \tag{3.19}$$

In general, if one has k parallel measurements on every subject where $j = 1, 2, \ldots, k$), then the expected value and variance of the sum of these measurements is given by

$$E\left(\sum_{j=1}^{k} X_{ij}\right) = k\tau_i \tag{3.20}$$

$$\text{Var}\left(\sum_{j=1}^{k} X_{ij}\right) = k^2\,\text{Var}(\tau_i) + k\,\text{Var}(e_{ij}) \tag{3.21}$$

Similarly,

$$\text{Var}(k\tau_i) = k^2\,\text{Var}(\tau_i) \tag{3.22}$$

The reliability of this sum is given by

$$R' = \frac{k^2\,\text{Var}(\tau_i)}{k^2\,\text{Var}(\tau_i) + k\,\text{Var}(e_{ij})} \tag{3.23}$$

$$= \frac{k^2 R}{k^2 R + k(1 - R)}$$

$$= \frac{kR}{kR + (1 - R)}$$

$$= \frac{kR}{1 + (k - 1)R} \tag{3.24}$$

Equation (3.24) is called the *Spearman–Brown prophesy formula.*

One thing that must be remembered in the derivation of reliability coefficients is that they are not a *fixed* characteristic of a test or measuring instrument. From Equation (3.9)

$$R = 1 - \frac{\text{Var}(e_{ij})}{\text{Var}(\tau_i) + \text{Var}(e_{ij})} \tag{3.25}$$

It should be clear from this form of the definition that R is dependent on the standard error of the test (that is, on $\text{Var}(e_{ij})$) and on the variability of the population being studied (that is, on $\text{Var}(\tau_i)$). For a given test it is assumed in classical test theory that the standard error of measurement is a constant characteristic of a measuring instrument. The variance of the true scores, on the other hand, changes from one population of subjects to another. The Wechsler deviation IQ, for example, has a standard deviation of 15 (variance, 225) when used in the general population. Its standard error of measurement is about 3 (variance, 9). The variance of true scores is about 216 ($225 - 9$). It follows that the reliability of a Wechsler deviation IQ is about 0.96 (216/225). If one were to use the Wechsler intelligence test on a population of mentally handicapped adults with a restricted range of IQ scores (standard deviation of, say, 8), then the reliability, would now be about 0.86 (($64 - 9$)/64).

Let X_{ij} and τ_i be observed and true values for measurements taken on the general population. Let the reliability be $R = \text{Var}(\tau_i)/\text{Var}(X_{ij})$. If measurements on a different population of subjects be X_{ij}^* and τ_i^* with reliability, $R^* = \text{Var}(\tau_i^*)/\text{Var}(X_{ij}^*)$, then, from Equation (3.9), it follows that

$$\text{Var}(e_{ij}) = \text{Var}(e_{ij}^*) = (1 - R)\,\text{Var}(X_{ij}) = (1 - R^*)\,\text{Var}(X_{ij}^*) \tag{3.26}$$

and

$$1 - R^* = (1 - R)\frac{\text{Var}(X_{ij})}{\text{Var}(X_{ij}^*)}$$

$$R^* = 1 - \frac{\text{Var}(X_{ij})}{\text{Var}(X_{ij}^*)}(1 - R) \tag{3.27}$$

Letting $\text{Var}(X_{ij}^*) = 225$, $\text{Var}(X_{ij}^*) = 64$ and $R = 0.96$, then it follows that

$$R^* = 1 - \frac{225}{64}(1 - 0.96)$$

$$= 0.86, \qquad \text{as before.}$$

3.2 Reliability of composite scores

Suppose one has obtained two measurements per subject, but in this case the two tests are not parallel (although they are assumed to be measuring the same trait). Here

$$X_{i1} = \tau_{i1} + e_{i1}$$

and

$$X_{i2} = \tau_{i2} + e_{i2}$$

Now consider the sum of these two scores:

$$X_i = X_{i1} + X_{i2}$$
$$= \tau_{i1} + \tau_{i2} + e_{i1} + e_{i2}$$
$$\text{Var}(X_i) = \text{Var}(X_{i1} + X_{i2})$$
$$= \text{Var}(X_{i1}) + \text{Var}(X_{i2}) + 2\,\text{Cov}(X_{i1}, X_{i2}) \qquad (3.28)$$
$$\text{Var}(\tau_{i1} + \tau_{i2}) = \text{Var}(\tau_{i1}) + \text{Var}(\tau_{i2}) + 2\,\text{Cov}(\tau_{i1}, \tau_{i2}) \qquad (3.29)$$

Now suppose that a common factor model holds (see Section 1.3) and

$$X_{i1} = \lambda_1 f_i + e_{i1}$$
$$X_{i2} = \lambda_2 f_i + e_{i2}$$

where f_i is the score of subject i on a latent trait or factor. Furthermore, let

$$\text{Var}(e_{i1}) = \sigma_1^2 \quad \text{and} \quad \text{Var}(e_{i2}) = \sigma_2^2$$

Then

$$X_i = (\lambda_1 + \lambda_2)f_i + e_{i1} + e_{i2}$$

and

$$\text{Var}(X_i) = (\lambda_1 + \lambda_2)^2\,\text{Var}(f_i) + \sigma_1^2 + \sigma_2^2 \qquad (3.30)$$

If, as is usual in factor analysis, an arbitrary constraint is introduced making $\text{Var}(f_i) = 1$, then

$$\text{Var}(X_i) = (\lambda_1 + \lambda_2)^2 + \sigma_1^2 + \sigma_2^2 \qquad (3.31)$$

and the variance of the subjects' true scores is given by

$$\text{Var}(\lambda_1 f_i + \lambda_2 f_i) = (\lambda_1 + \lambda_2)^2 \qquad (3.32)$$

The reliability of the composite score, X_i, is given by

$$R = \frac{(\lambda_1 + \lambda_2)^2}{(\lambda_1 + \lambda_2)^2 + \sigma_1^2 + \sigma_2^2} \qquad (3.33)$$

A more interesting problem is to estimate the reliability of a weighted sum of X_{i1} and X_{i2} and to select weights so that the reliability of this composite score is maximized. The weighted average is given by

$$X_i = W_1 X_{i1} + W_2 X_{i2}$$

where

$$W_1 + W_2 = 1$$

Still constraining $\text{Var}(f_i)$ to be equal to 1, it follows that

$$\text{Var}(X_i) = (W_1\lambda_1 + W_2\lambda_2)^2 + W_1^2\sigma_1^2 + W_2^2\sigma_2^2 \qquad (3.34)$$

and the reliability of the composite is

$$R = \frac{(W_1\lambda_1 + W_2\lambda_2)^2}{(W_1\lambda_1 + W_2\lambda_2)^2 + W_1^2\sigma_1^2 + W_2^2\sigma_1^2} \qquad (3.35)$$

and, since $W_2 = 1 - W_1$

$$R = \frac{(W_1\lambda_1 + (1 - W_1)\lambda_2)^2}{(W_1\lambda_1 + (1 - W_1)\lambda_2)^2 + W_1^2\sigma_1^2 + (1 - W_1)^2\sigma_2^2} \qquad (3.36)$$

Taking logarithms:

$$\log R = 2\log(W_1\lambda_1 + (1 - W_1)\lambda_2)$$
$$- \log[(W_1\lambda_1 + (1 - W_1)\lambda_2)^2 + W_1^2\sigma_1^2 + (1 - W_1)^2\sigma_2^2]$$

Differentiating with respect to W_1:

$$\frac{d\log R}{dW_1} = \frac{2(\lambda_1 - \lambda_2)}{W_1\lambda_1 + (1 - W_1)\lambda_2}$$

$$- \frac{(W_1\lambda_1 + (1 - W_1)\lambda_2)(\lambda_1 - \lambda_2) + 2W_1\sigma_1^2 - 2(1 - W_1)\sigma_2^2}{[W_1\lambda_1 + (1 - W_1)\lambda_2]^2 + W_1^2\sigma_1^2 + (1 - W_2)^2\sigma_2^2}$$

When R is at a maximum this derivative is equal to zero and it follows that

$$W_1 = \frac{\lambda_1/\sigma_1^2}{\lambda_1/\sigma_1^2 + \lambda_2/\sigma_2^2} \qquad (3.37)$$

It will be left to the reader to derive (3.37) and to show that this is a solution for a maximum value of R, rather than a minimum. Obviously

$$W_2 = \frac{\lambda_2/\sigma_2^2}{\lambda_1/\sigma_1^2 + \lambda_2/\sigma_2^2}$$

In the general case of a weighted sum of k congeneric tests the reliability is maximum when the weight for the jth test is given by

$$W_j \propto \frac{\lambda_j}{\sigma_j^2} \qquad (3.38)$$

If these tests are, in fact, essentially τ-equivalent (the λs are all the same), then

$$W_j \propto \frac{1}{\sigma_j^2} \qquad (3.39)$$

3.3 Split halves and the internal consistency of tests

One characteristic of many behavioural and social measurements that distinguishes them from physical measurements is that they are obtained from the response to several different questions or test items. The weight of an object is given by a single instrument reading whereas an intelligence quotient may be calculated from answers to fifty or a hundred individual tests (items) of cognitive ability. One traditional way of obtaining the reliability of a psychological measurement is to split the test into two equal sized groups of comparable items and to estimate the reliability of the total test score from the subtotals obtained from the two halves. Let the subtotal obtained by subject i on the first half be represented by X_{i1}, and, similarly, let the corresponding subtotal for the second half be X_{i2}. If

$$X_{i1} = \tau_{i1} + e_{i1}$$

and

$$X_{i2} = \tau_{i2} + e_{i2}$$

where $E(X_{i1}) = \tau_{i1}$ and $E(X_{i2}) = \tau_{i2}$ as before, then the reliability of the sum of these subtotals

$$X_i = X_{i1} + X_{i2}$$

is given by

$$R = \frac{\text{Var}(\tau_{i1} + \tau_{i2})}{\text{Var}(X_{i1} + X_{i2})}$$

or

$$R = 1 - \frac{\text{Var}(e_{i1}) + \text{Var}(e_{i2})}{\text{Var}(X_{i1} + X_{i2})}$$

Now, if one is prepared to assume that the two split halves are essentially τ-equivalent (but not necessarily parallel), then

$$\text{Var}(e_{i1}) + \text{Var}(e_{i2}) = \text{Var}(X_{i1} + X_{i2}) - 2[\text{Var}(X_{i1}) + \text{Var}(X_{i2})]$$

It follows that

$$R = 2\left[1 - \frac{\text{Var}(X_{i1}) + \text{Var}(X_{i2})}{\text{Var}(X_{i1} + X_{i2})}\right] \tag{3.40}$$

An alternative expression is

$$R = \frac{4\,\text{Cov}(X_{i1}, X_{i2})}{\text{Var}(X_{i1} + X_{i2})} \tag{3.41}$$

The major problem with the split-half reliability coefficient as defined by Equation (3.40) is that for a given test it is not unique. There are $(2N)!/2(N!)^2$ possible splits of a test dividing $2N$ items into two equal halves containing N items. Let the average of $\text{Cov}(X_{i1}, X_{i2})$ over all these possible splits be represented by the expected value $E[\text{Cov}(X_{i1}, X_{i2})]$; then the mean split-half reliability coefficient over all possible splits is given by

$$E(R) = \frac{4E[\text{Cov}(X_{i1}, X_{i2})]}{\text{Var}(X_{i1} + X_{i2})} \tag{3.42}$$

where $E[R]$ is again the expectation over all possible splits. (Note that $\text{Var}(X_{i1} + X_{i2})$ is fixed for all possible splits of the test.)

Let the $2N$ item scores for individual i be represented by $Z_{i1}, Z_{i2}, ..., Z_{iN}$, where, for a given arbitrary split, it is assumed that the first N item scores be added to give X_{i1} and the second N items are added to give X_{i2}. Then $\text{Cov}(X_{i1}, X_{i2})$ is given by

$$\text{Cov}(X_{i1}, X_{i2}) = \sum_{j=1}^{N} \sum_{k=N+1}^{2N} \text{Cov}(Z_{ij}, Z_{ik}) \tag{3.43}$$

It follows that, over all possible splits,

$$E(\text{Cov}(X_{i1}, X_{i2})) = \sum_{j=1}^{N} \sum_{k=N+1}^{2N} E[\text{Cov}(Z_{ij}, Z_{ik})] \tag{3.44}$$

Note that there are N^2 components in the sum on the right of Equation (3.44). For any particular split all pairs of items have an equal chance of being assigned to the labels Z_{ij} $(j = 1, 2, ..., N)$ and Z_{ik} $(k = N+1, N+2, ..., 2N)$. Therefore all the $E[\text{Cov}(Z_{ij}, Z_{ik})]$ terms in the double sum of Equation (3.44) are equal and, in addition, equal to the mean covariance between all distinct items of the test. That is

$$E[\text{Cov}(Z_{ij}, Z_{ik})] = \frac{\sum_{s=1}^{2N} \sum_{t=1, t>s}^{2N} \text{Cov}(Z_{is}, Z_{it})}{N(2N-1)} \tag{3.45}$$

where, in this case, the labelling of the items Z_{is} and Z_{it} is such that both s and t can take any value from 1 to $2N$. It follows from Equations (3.42), (3.44) and (3.45) that

$$E(R) = \frac{N^2}{\text{Var}(X_i)} \left[\frac{2 \sum_{s=1}^{2N} \sum_{t=1, t>s}^{2N} \text{Cov}(Z_{is}, Z_{it})}{N(2N-1)} \right]$$

$$= \frac{2N}{(2N-1)\,\text{Var}(X_i)} \left[\text{Var}(X_i) - \sum_{j=1}^{2N} \text{Var}(Z_{ij}) \right]$$

$$= \left(\frac{2N}{2N-1} \right) \left(1 - \frac{\sum_{j=1}^{2N} \text{Var}(Z_{ij})}{\text{Var}(X_i)} \right) \tag{3.46}$$

The more familiar form of Equation (3.46) is defined for a test of N items, that is

$$E(R) = \frac{N}{N-1} \left(1 - \frac{\sum_{j=1}^{N} \text{Var}(Z_{ij})}{\text{Var}(X_i)} \right) \tag{3.47}$$

This is usually referred to as *Cronbach's α coefficient* (Cronbach, 1951; Lord and Novick, 1968, Chapter 4). It can be thought of as a measure of the *internal consistency* of a psychometric test. A particular case of interest arises when the items are all binary. In this case $\text{Var}(Z_{ij}) = P_j(1 - P_j)$, where P_j is the probability of a positive response to item j, and Equation (3.47) becomes

$$E(R) = \frac{N}{N-1} \left(1 - \frac{\sum_{j=1}^{N} P_j(1 - P_j)}{\text{Var}(X_i)} \right) \tag{3.48}$$

Here $E(R)$ is known as the Kuder–Richardson formula 20 or KR20 (Kuder and Richardson, 1937).

Now, the derivation of the split-half reliability coefficients and both Cronbach's α and KR20 is dependent on the assumption that the split halves are essentially τ-equivalent. For Cronbach's α and KR20 this is equivalent to the assumption that all of the individual items are essentially τ-equivalent too. Of course, in terms of its mathematical properties, it makes no difference whether the composite score X is the sum of N item scores or of N subtest scores, $X_1, X_2, \ldots, X_N$. Let $X_i = X_{i1} + X_{i2} + \cdots + X_{iN}$. Similarly, let $\tau_i = \tau_{i1} + \tau_{i2}, \ldots, \tau_{iN}$. As before, the subscript i refers to measurements on the ith subject. Then

$$\alpha = \frac{N}{N-1}\left[1 - \frac{\sum_{j=1}^{N}\mathrm{Var}(X_{ij})}{\mathrm{Var}(X_i)}\right] \tag{3.49}$$

The actual reliability of X is given by

$$R = \frac{\mathrm{Var}(\tau_i)}{\mathrm{Var}(X_i)}$$

and it will now be demonstrated that

$$R \geq \alpha$$

That is, α provides a lower bound for the true reliability, R.

The proof of the above inequality starts with the inequality

$$[\mathrm{Var}(\tau_{ij}) - \mathrm{Var}(\tau_{ik})]^2 \geq 0 \tag{3.50}$$

or

$$\mathrm{Var}(\tau_{ij}) + \mathrm{Var}(\tau_{ik}) \geq 2\,\mathrm{sd}(\tau_{ij})\,\mathrm{sd}(\tau_{ik}) \tag{3.51}$$

Furthermore,

$$\frac{\mathrm{Cov}(\tau_{ij}, \tau_{ik})}{\mathrm{sd}(\tau_{ij})\,\mathrm{sd}(\tau_{ik})} \leq \frac{|\mathrm{Cov}(\tau_{ij}, \tau_{ik})|}{\mathrm{sd}(\tau_{ij})\,\mathrm{sd}(\tau_{ik})} \leq 1 \tag{3.52}$$

and, therefore,

$$\mathrm{sd}(\tau_{ij})\,\mathrm{sd}(\tau_{ik}) \geq |\mathrm{Cov}(\tau_{ij}, \tau_{ik})| \geq \mathrm{Cov}(\tau_{ij}, \tau_{ik}) \tag{3.53}$$

It follows from Equations (3.51) and (3.53) that

$$\mathrm{Var}(\tau_{ij}) + \mathrm{Var}(\tau_{ik}) \geq 2\,\mathrm{Cov}(\tau_{ij}, \tau_{ik}) \tag{3.54}$$

Note that the assumption that the N tests are congeneric has not been made in the derivation of (3.54). If this assumption were to be made then (3.54) would still hold, of course, but Equation (3.52) could be replaced by

$$\frac{\mathrm{Cov}(\tau_{ij}, \tau_{ik})}{\mathrm{sd}(\tau_{ij})\,\mathrm{sd}(\tau_{ik})} = 1$$

Summing over all component tests j and k (with $j \neq k$) it follows that

$$\sum_{\substack{j=1 \\ k \neq j}}^{N}\sum_{k=1}^{N}[\mathrm{Var}(\tau_{ij}) + \mathrm{Var}(\tau_{ik})] \geq 2\sum_{\substack{j=1 \\ k \neq j}}^{N}\sum_{k=1}^{N}\mathrm{Cov}(\tau_{ij}, \tau_{ik}) \tag{3.55}$$

Now, letting $j = k$ in the summation

$$\sum_{j=1}^{N} \sum_{k=1}^{N} [\text{Var}(\tau_{ij}) + \text{Var}(\tau_{ik})] = \sum_{j=1}^{N} \left[N\,\text{Var}(\tau_{ij}) + \sum_{k=1}^{N} \text{Var}(\tau_{ik}) \right]$$

$$= N\sum_{j=1}^{N} \text{Var}(\tau_{ij}) + N\sum_{k=1}^{N} \text{Var}(\tau_{ik})$$

$$= 2N\sum_{j=1}^{N} \text{Var}(\tau_{ij}) \tag{3.56}$$

Also

$$\sum_{j=1}^{N} \sum_{k=1}^{N} [\text{Var}(\tau_{ij}) + \text{Var}(\tau_{ik})]$$

$$= \sum_{\substack{j=1 \\ }}^{N} \sum_{\substack{k=1 \\ k \neq j}}^{N} [\text{Var}(\tau_{ij}) + \text{Var}(\tau_{ik})] + \sum_{\substack{j=1 \\ }}^{N} \sum_{\substack{k=1 \\ k = j}}^{N} [\text{Var}(\tau_{ij}) + \text{Var}(\tau_{ik})]$$

$$= \sum_{\substack{j=1 \\ }}^{N} \sum_{\substack{k=1 \\ k \neq j}}^{N} [\text{Var}(\tau_{ij}) + \text{Var}(\tau_{ik})] + 2\sum_{j=1}^{N} \text{Var}(\tau_{ij}) \tag{3.57}$$

Therefore, from (3.56) and (3.57)

$$\sum_{\substack{j=1 \\ }}^{N} \sum_{\substack{k=1 \\ k \neq j}}^{N} [\text{Var}(\tau_{ij}) + \text{Var}(\tau_{ik})] = 2N\sum_{j=1}^{N} \text{Var}(\tau_{ij}) - 2\sum_{j=1}^{N} \text{Var}(\tau_{ij})$$

$$= 2(N-1)\sum_{j=1}^{N} \text{Var}(\tau_{ij}) \tag{3.58}$$

The inequality expressed in Equation (3.55) is equivalent to

$$2(N-1)\sum_{j=1}^{N} \text{Var}(\tau_{ij}) \geq 2\sum_{\substack{j=1 \\ }}^{N} \sum_{\substack{k=1 \\ k \neq j}}^{N} \text{Cov}(\tau_{ij}, \tau_{ik})$$

$$\sum_{j=1}^{N} \text{Var}(\tau_{ij}) \geq \frac{\sum_{j=1}^{N} \sum_{k=1, k \neq j}^{N} \text{Cov}(\tau_{ij}, \tau_{ik})}{N-1} \tag{3.59}$$

Now

$$\text{Var}(\tau_i) = \sum_{j=1}^{N} \text{Var}(\tau_i) + \sum_{\substack{j=1 \\ }}^{N} \sum_{\substack{k=1 \\ k \neq j}}^{N} \text{Cov}(\tau_{ij}, \tau_{ik})$$

So, from (3.59)

$$\text{Var}(\tau_i) \geq \frac{\sum_{j=1}^{N} \sum_{k=1, k \neq j}^{N} \text{Cov}(\tau_{ij}, \tau_{ik})}{N-1} + \sum_{\substack{j=1 \\ }}^{N} \sum_{\substack{k=1 \\ k \neq j}}^{N} \text{Cov}(\tau_{ij}, \tau_{ik})$$

or

$$\text{Var}(\tau_i) \geq \frac{N}{N-1} \sum_{\substack{j=1 \\ }}^{N} \sum_{\substack{k=1 \\ k \neq j}}^{N} \text{Cov}(\tau_{ij}, \tau_{ik}) \tag{3.60}$$

But

$$\text{Var}(X_i) - \sum_{j=1}^{N} \text{Var}(X_{ij}) = \sum_{j=1}^{N} \sum_{\substack{k=1 \\ k \neq j}}^{N} \text{Cov}(X_{ij}, X_{ik})$$

$$= \sum_{j=1}^{N} \sum_{\substack{k=1 \\ k \neq j}}^{N} \text{Cov}(\tau_{ij}, \tau_{ik}) \qquad (3.61)$$

Substituting (3.60) into (3.61), it follows that

$$\text{Var}(\tau_i) \geq \frac{N}{N-1} \left[\text{Var}(X_i) - \sum_{j=1}^{N} \text{Var}(X_{ij}) \right]$$

Dividing by $\text{Var}(X_i)$ gives the required result.

$$R = \frac{\text{Var}(\tau_i)}{\text{Var}(X_i)} \geq \frac{N}{N-1} \left[1 - \frac{\sum_{j=1}^{N} \text{Var}(X_{ij})}{\text{Var}(X_i)} \right] \qquad (3.62)$$

In recent years there have been several other internal consistency coefficients proposed as alternatives to Cronbach's α. These coefficients include θ (Armor, 1974), ω (Heise and Bohrnstedt, 1970) and the α_0 coefficient of Bentler (1968). These will not be described here. The interested reader is referred to detailed discussions in K. W. Smith (1974) and in Greene and Carmines (1979).

3.4   Repeated testing and stability

So far the discussion of reliability has centred on the comparison of alternative forms of a measuring instrument or psychometric test, or on the internal consistency of psychometric tests. Another traditional approach is based on the idea of repeating the measurements using the same test. An instrument is judged to be reliable if there is close agreement between the measurements made at the two times. The data given in Table 1.1 is an example for physical measurements such as the weights of packets of potatoes. When one considers psychometric tests there is a problem that is not usually encountered in the case of physical weights and that is that there is a possibility that the trait or characteristic being measured may have actually changed. There is also a problem caused by learning and memory, but for the time being this will be ignored. If for example, one has a questionnaire designed to assess anxiety, for example, it is highly likely that a subject's 'true' level of anxiety will change with time. The lack of agreement between measures of anxiety taken at two different times could be caused by either changes on the subject's true state or measurement error, or both. In this case stability of the trait or characteristic being measured will be confounded with test reliability. Of course, if an investigator is convinced that a stable trait is in fact being measured, then this problem might safely be ignored. The reliability of the test could then be assessed as if the initial test scores and the re-test scores were obtained from alternative test forms (which may, or may not, be assumed to be parallel) as has been described above.

Returning to the example of the measurement of anxiety, if one wishes to disentangle the effects of lack of stability from the effects of poor instrument precision, then more data are needed. As was pointed out in Section 1.2 two measurements per subject are not sufficient to be able or to make inferences about variation in *both* true scores and errors and in the example of the weights of potatoes an assumption was made that, at most, a *constant* difference in the true weights at the two true points exist. In the case of anxiety this assumption might not be justified.

An early attempt at separating reliability and stability was made by Heise (1969). A more realistic solution was provided by Wiley and Wiley (1970), and it is this that will form the basis for the present discussion. For subject i assume that there have been three separate measurements made (that is, at three consecutive times). These measurements are X_{i1}, X_{i2} and X_{i3}. The relationships between these observed measurements and the subject's true scores are given by the following:

$$X_{i1} = \tau_{i1} + e_{i1}$$
$$X_{i2} = \tau_{i2} + e_{i2} \qquad (3.63)$$
$$X_{i3} = \tau_{i3} + e_{i3}$$

where τ_{i1}, τ_{i2} and τ_{i3} are the true scores for the three time points, respectively, and similarly e_{i1}, e_{i2} and e_{i3} are the three corresponding measurement errors. The measurement errors are assumed to be uncorrelated with each other and with each of the true scores. Now, the Wiley model also involves specifying a set of *linear structural equations* describing the assumed relationships between the three true scores. These are

$$\tau_{i1} = \theta_{i1}$$
$$\tau_{i2} = \beta_1 \theta_{i1} + \theta_{i2} \qquad (3.64)$$

and

$$\tau_{i3} = \beta_2(\beta_1 \theta_{i1} + \theta_{i2}) + \theta_{i3}$$

Substitution of the expressions for the true scores in (3.64) into the measurement model described in Equations (3.63) yields a new set of equations as follows:

$$X_{i1} = \theta_{i1} + e_{i1}$$
$$X_{i2} = \beta_1 \theta_{i1} + \theta_{i2} + e_{i2} \qquad (3.65)$$
$$X_{i3} = \beta_2(\beta_1 \theta_{i1} + \theta_{i2}) + \theta_{i3} + e_{i3}$$

Basically the Wiley model is postulating that a subject's true score at the second testing is linearly related to, but not necessarily the same as, the true score at the first testing. Similarly the true score at the third testing is linearly related to that at the second testing. The partial correlation between the true scores at the third and first testing, given the true score at the second, is, however, assumed to be zero.

Using the usual definition of reliability, the reliabilities of the test at the three times are given by

$$R_1 = \frac{\text{Var}(\theta_{i1})}{\text{Var}(\theta_{i1}) + \text{Var}(e_{i1})}$$

$$R_2 = \frac{\beta_1^2 \,\text{Var}(\theta_{i1}) + \text{Var}(\theta_{i2})}{\beta_1^2 \,\text{Var}(\theta_{i1}) + \text{Var}(\theta_{i2}) + \text{Var}(e_{i2})}$$

$$R_3 = \frac{\beta_2^2[\beta_1^2 \,\text{Var}(\theta_{i1}) + \text{Var}(\theta_{i2})] + \text{Var}(\theta_{i3})}{\beta_2^2[\beta_1^2 \,\text{Var}(\theta_{i1}) + \text{Var}(\theta_{i2})] + \text{Var}(\theta_{i3}) + \text{Var}(e_{i3})}$$

$$(3.66)$$

If one further assumption is made, that is

$$\text{Var}(e_{i1}) = \text{Var}(e_{i2}) = \text{Var}(e_{i3}) = \text{Var}(e_i)$$

then it can be seen from Equations (3.66) that the reliability of the test changes with time. Reliability is not a fixed characteristic of a test. If one was prepared to let the variances of the error terms change, however, then one could equate the three reliabilities given by Equations (3.66). This does not appear to be a particularly sensible suggestion, however. Wiley and Wiley (1970) defined the *stability* coefficient γ_{jk} as the correlation between true scores at times j and k. The three stability coefficients are then given as follows:

$$\gamma_{12} = \beta_1 \frac{\sqrt{(\text{Var}(\theta_{i1}))}}{\sqrt{(\beta_1^2 \,\text{Var}(\theta_{i1}) + \text{Var}(\theta_{i2}))}}$$

$$\gamma_{23} = \frac{\beta_2 \sqrt{(\beta_1^2 \,\text{Var}(\theta_{i1}) + \text{Var}(\theta_{i2}))}}{\sqrt{(\beta_2^2(\beta_1^2 \,\text{Var}(\theta_{i1}) + \text{Var}(\theta_{i2})) + \text{Var}(\theta_{i3}))}}$$

$$(3.67)$$

$$\gamma_{13} = \beta_1 \beta_2 \frac{\sqrt{(\text{Var}(\theta_{i1}))}}{\sqrt{(\beta_2^2(\beta_1^2 \,\text{Var}(\theta_{i1}) + \text{Var}(\theta_{i2})) + \text{Var}(\theta_{i3}))}}$$

Note that there are six parameters to be estimated from a given set of data (the variance–covariance matrix for the three measurements). The model is *just identified* (see Chapter 5). The six parameters to be estimated are $\text{Var}(\theta_{i1})$, $\text{Var}(\theta_{i2})$ and $\text{Var}(\theta_{i3})$, β_1 and β_2, and $\text{Var}(e_i)$. The six items of data from which they can be estimated are $\text{Var}(X_{i1})$, $\text{Var}(X_{i2})$, $\text{Var}(X_{i3})$, $\text{Cov}(X_{i1}, X_{i2})$, $\text{Cov}(X_{i1}, X_{i3})$, and $\text{Cov}(X_{i2}, X_{i3})$. The expected values for these variances and covariances are shown in Table 3.1(a), and the corresponding parameter estimates obtained from equating observed and expected values for the variances and covariances is shown in Table 3.1(b).

Wiley and Wiley (1970) illustrated their model through the analysis of a covariance matrix for subjects' reported earnings for three successive years. These covariances are based on the reported earnings from a sample of 6 222 men for the years 1962–64. The covariances and corresponding correlations

Table 3.1

(a) Expected covariance matrix for Wiley and Wiley model

$$\mathrm{Var}(X_{i1}) = \mathrm{Var}(\theta_{i1}) + \mathrm{Var}(e_i)$$
$$\mathrm{Cov}(X_{i1}, X_{i2}) = \beta_1 \mathrm{Var}(\theta_{i1})$$
$$\mathrm{Cov}(X_{i1}, X_{i3}) = \beta_1 \beta_2 \mathrm{Var}(\theta_{i1})$$
$$\mathrm{Var}(X_{i2}) = \beta_1^2 \mathrm{Var}(\theta_{i1}) + \mathrm{Var}(\theta_{i2}) + \mathrm{Var}(e_i)$$
$$\mathrm{Cov}(X_{i2}, X_{i3}) = \beta_2[\beta_1^2 \mathrm{Var}(\theta_{i1}) + \mathrm{Var}(\theta_{i2})]$$
$$\mathrm{Var}(X_{i3}) = \beta_2^2[\beta_1^2 \mathrm{Var}(\theta_{i1}) + \mathrm{Var}(\theta_{i2})] + \mathrm{Var}(\theta_{i3}) + \mathrm{Var}(e_i)$$

(b) Parameters and their estimators for the Wiley and Wiley Model

Parameter	Estimator
β_2	$\widehat{\mathrm{Cov}}(X_{i1}, X_{i3})/\widehat{\mathrm{Cov}}(X_{i1}, X_{i2})$
$\mathrm{Var}(e_i)$	$\widehat{\mathrm{Var}}(X_{i2}) - [\widehat{\mathrm{Cov}}(X_{i2}, X_{i3})/\hat{\beta}_2]$
$\mathrm{Var}(\theta_{i1})$	$\widehat{\mathrm{Var}}(X_{i1}) - \widehat{\mathrm{Var}}(e_i)$
β_1	$\widehat{\mathrm{Cov}}(X_{i1}, X_{i2})/\widehat{\mathrm{Var}}(\theta_{i1})$
$\mathrm{Var}(\theta_{i2})$	$\widehat{\mathrm{Var}}(X_{i2}) - [\hat{\beta}_1 \widehat{\mathrm{Cov}}(X_{i1}, X_{i2}) + \widehat{\mathrm{Var}}(e_i)]$
$\mathrm{Var}(\theta_{i3})$	$\widehat{\mathrm{Var}}(X_{i3}) - [\hat{\beta}_2 \widehat{\mathrm{Cov}}(X_{i2}, X_{i3}) + \widehat{\mathrm{Var}}(e_i)]$

are shown in Table 3.2. Estimates of the six parameters of the model are the following:

$$\hat{\beta}_2 = 1.035, \qquad \hat{\beta}_1 = 1.019$$

$$\widehat{\mathrm{Var}}(e_i) = 1.581 \times 10^6$$

$$\widehat{\mathrm{Var}}(\theta_{i1}) = 9.914 \times 10^6$$

$$\widehat{\mathrm{Var}}(\theta_{i2}) = 1.116 \times 10^6$$

$$\widehat{\mathrm{Var}}(\theta_{i3}) = 1.906 \times 10^6$$

The estimated reliability and stability coefficients are

$$\hat{R}_1 = 0.862 \qquad \hat{\gamma}_{12} = 0.950$$

$$\hat{R}_2 = 0.878 \qquad \hat{\gamma}_{23} = 0.930$$

$$\hat{R}_3 = 0.899 \qquad \hat{\gamma}_{13} = 0.884$$

Table 3.2 Sample covariances* (and correlations in parentheses) for reported earnings in three successive years

	X_1	X_2	X_3
X_1: 1962 earnings	11.495		
X_2: 1963 earnings	10.106 (0.827)	12.995	
X_3: 1964 earnings	10.455 (0.778)	11.808 (0.827)	15.708

* $\times 10^{-6}$. Data from Cutright, as reported by Wiley and Wiley (1970).

3.5   Reliability of counts and proportions

Poisson process models have been introduced in Section 1.4 where it was shown that the standard error of measurement for a count with an expected value of τ_i is $\sqrt{\tau_i}$. This was based on the two important properties of the Poisson distribution described by Equation (1.21); that is,

$$E(X_i \mid \tau_i) = \tau_i$$

$$\mathrm{Var}(X_i \mid \tau_i) = \tau_i$$

Here X_i is the observed count with Poisson parameter τ_i. The subscript i indicates the ith subject. If one now considers a population of subjects, then one can define the following parameters:

$$E(\tau_i) = \mu_\tau$$

$$\mathrm{Var}(\tau_i) = \sigma_\tau^2$$

The reliability of the test or measurement procedure that is generating these counts can then be defined by

$$R = \frac{\mathrm{Var}(\tau_i)}{\mathrm{Var}(X_i)}$$

where the variance of the counts ($\mathrm{Var}(X_i)$) is over the whole population (that is, not conditional on τ_i). X_i is distributed according to a *compound Poisson distribution*.

It is straightforward to derive the general result for a compound distribution

$$\mathrm{Var}(X_i) = E[\mathrm{Var}(X_i \mid \tau_i)] + \mathrm{Var}[E(X_i \mid \tau_i)] \tag{3.68}$$

(see Allison, 1978 and Parzen, 1962), which, in the case of the compound Poisson distribution, reduces to

$$\mathrm{Var}(X_i) = E[\tau_i] + \mathrm{Var}[\tau_i]$$

$$= \mu_\tau + \sigma_\tau^2 \tag{3.69}$$

Therefore

$$\sigma_\tau^2 = \mathrm{Var}(X_i) - E(X_i)$$

where, of course, $E(X_i) = \mu_\tau$. Substituting Equation (3.69) into the reliability definition yields

$$R = \frac{\mathrm{Var}(X_i) - E(X_i)}{\mathrm{Var}(X_i)}$$

$$= 1 - \frac{E(X_i)}{\mathrm{Var}(X_i)} \tag{3.70}$$

This reliability can be estimated from the observed mean and variance of the sample counts.

One can use a similar argument to derive a reliability coefficient for the number-right score from a test of N binary responses (see Section 1.5). Let X_i be the number-right score for subject i and let X_i follow a binomial distribution with parameter P_i. Then it follows that

$$E(X_i | P_i) = Np_i = \tau_i \tag{3.71}$$

$$\text{Var}(X_i | P_i) = Np_i(1 - P_i) = \tau_i\left(1 - \frac{\tau_i}{N}\right) \tag{3.72}$$

where, as before, τ_i is defined to be the true score for the ith subject. Again, considering the whole population of subjects one defines

$$E(\tau_i) = \mu_\tau$$

and

$$\text{Var}(\tau_i) = \sigma_\tau^2$$

The count X_i can be considered to be distributed according to a *compound binomial distribution* with variance given by substituting (3.71) and (3.72) into Equation (3.68); that is

$$\text{Var}(X_i) = E\left[\tau_i\left(1 - \frac{\tau_i}{N}\right)\right] + \text{Var}(\tau_i)$$

$$= E(\tau_i) - \frac{E(\tau_i^2)}{N} + \sigma_\tau^2$$

$$= \mu_\tau - \frac{1}{N}[\sigma_\tau^2 + \mu_\tau^2] + \sigma_\tau^2$$

$$= \mu_\tau - \frac{\mu_\tau^2}{N} + \frac{(N-1)}{N}\sigma_\tau^2$$

Therefore

$$\sigma_\tau^2 = \frac{N}{N-1}\left[\text{Var}(X_i) - \mu_\tau + \frac{\mu_\tau^2}{N}\right]$$

If the reliability of the test is defined in the usual way, then it follows that

$$R = \frac{N}{N-1}\left[1 - \frac{\mu_\tau[1 - (\mu_\tau/N)]}{\text{Var}(X_i)}\right]$$

or, if $\mu_p = \mu_\tau/N$,

$$R = \frac{N}{N-1}\left[1 - \frac{N\mu_p(1 - \mu_p)}{\text{Var}(X_i)}\right] \tag{3.73}$$

This is equivalent to formula 21 (that is, KR21) of Kuder and Richardson (1937). It can be derived from the internal consistency formula KR20 (see Equation (3.48)) if $P_j = \mu_p$ for all items. The latter does not necessarily imply equal item difficulties, but can arise, for example, if one considers the items actually used to be randomly sampled from a much larger population of possible items. Here the probability of the ith subject currently answering an item selected at random is assumed to be P_i.

3.6   Generalisability theory

An investigator who takes a measurement twice is likely to obtain results that differ. This may be due to changes in the measuring instrument, the conditions of measurement or fluctuations in the underlying trait or characteristic that is being measured. In classical test theory an observation is assumed to be a combination of an individual's true score (defined as an expected value) and a random measurement error. The sources of variation in a subject's scores or measurements are not explicitly allowed for in classical test theory, although the investigation of sets of congeneric or τ-equivalent tests does enable one to study relative biases. In a *generalisability study* (Cronbach *et al.*, 1972) one sets out to systematically investigate the sources of variation of measurements. A behavioural measurement is assumed to be a sample from a *universe of admissible observations*, characterised by one or more *facets*. A facet (or experimental factor) can include, for example, alternative measuring instruments or test forms, different occasions, different clinics or laboratories, or different clinicians or scientists within clinics or laboratories, respectively. The different *levels* of a facet (such as alternative test forms) are called measurement or test conditions. For a given generalisability study one defines a universe of admissible observations by listing the measurement condition for each of the facets of interest. Consider, for example, a generalisability study in which three raters each assess subjects' anxiety in the morning and again in the evening. This is an example of a two-facet design. The first facet (rater) has three levels; the second (time of day) has two. In all there are six combinations of measurement conditions. A subject's *universe score* is considered to be the expected value of the score over the universe of admissible observations. A different universe of admissible observations (different rating scales, for example) would imply a different universe score. A decision maker who, in the future, wishes to apply the same rating or measurement technique, wishes to generalize to some universe of conditions, all of which can be considered to be providing samples of the same information. This is referred to as the *universe of generalisation*. It may be identical to the universe of admissible observations, but not necessarily so. The decision maker, for example, may decide to use only one of the above raters for further measurements or measurements might only ever be taken in the morning. This choice would, of course, imply a change in a subject's universe score.

Returning to the two-facet generalisability study, it should be clear that raters and time of rating contribute to variation in subjects' scores in addition to undifferentiated measurement error (random fluctuations). Decision makers' ratings, for example, are likely to be much more precise or consistent (reliable) if they are made either at one fixed time of day or by one individual rater, or both. Similarly, a decision maker is likely to get a more precise rating if she were to use an average rating obtained from two or more of the admissible conditions.

To clarify and extend some of these concepts, consider a simple single facet design involving each subject being rated by each of *J* raters. Both subjects and raters are thought of as random samples from large populations of possible subjects and populations of possible raters, respectively. An observation made by the jth rater ($j = 1, 2, ..., J$) on the ith subject ($i = 1, 2, ..., I$) can be represented by

$$X_{ij} = \mu + \alpha_i + \beta_j + e_{ij} \tag{3.74}$$

observed general subject rater residual
rating mean effect effect

If subjects are only rated once by each of the three raters, then e_{ij} will be a confounded combination of the subject–rater interaction and random measurement error. In addition

$$\mu_i = \underset{j}{E}(X_{ij}) = \mu + \alpha_i \tag{3.75}$$

where μ_i is the ith subject's universe score

$$\underset{i}{E}(X_{ij}) = \mu + \beta_j \tag{3.76}$$

and

$$\underset{i,j}{E}(X_{ij}) = \mu \tag{3.77}$$

$$\begin{aligned}
\text{Var}(X_{ij}) &= \underset{i,j}{E}(X_{ij} - \mu)^2 \\
&= \underset{j}{E}(\alpha_i^2) + \underset{i}{E}(\beta_j^2) + \underset{i,j}{E}(e_{ij}^2) \\
&= \text{Var}(\alpha_i) + \text{Var}(\beta_j) + \text{Var}(e_{ij})
\end{aligned} \tag{3.78}$$

$$\text{Var}(\mu_i) = \text{Var}(\alpha_i) \tag{3.79}$$

One can use analysis of variance programs to estimate each of these components of variance using data collected in a given generalisability study (see Chapter 6).

For the single-facet generalisability study one example of a reliability or *generalisability* coefficient is defined by

$$R = \frac{\text{Var}(\mu_i)}{\text{Var}(X_{ij})} \tag{3.80}$$

$$= \frac{\text{Var}(\alpha_i)}{\text{Var}(\alpha_i) + \text{Var}(\beta_j) + \text{Var}(e_{ij})} \tag{3.81}$$

This corresponds to the reliability of single measurements made on each subject where the rater for each subject is randomly selected for the population of possible raters. If there are no relative biases between raters (that is, $\text{Var}(\beta_j) = 0$) or, alternatively, only *one* rater is to be used for assessing *all* subjects in a future study, then the reliability or generalisability coefficient for these measurements would be

$$R = \frac{\text{Var}(\alpha_i)}{\text{Var}(\alpha_i) + \text{Var}(e_{ij})} \tag{3.82}$$

Equation (3.82) would provide a realistic measure of reliability only if $\text{Var}(e_{ij})$ were solely derived from variation due to random fluctuations (that is, there were no subject–rater interaction), otherwise it would underestimate the true reliability. Reliability coefficients derived from components of variance will be discussed in greater detail in Chapter 6. It is necessary only to point out here that the particular form of a reliability or generalisability coefficient will depend on the way a measurement is to be used in a given investigation.

3.7 Exercises

3.1   Calculate Cronbach's coefficient for
(a) The two measuring instruments in Table 1.1.
(b) The string lengths provided by Graham, Brian and David in Table 1.2.

3.2   Calculate the internal consistency coefficients KR20 and KR21 from the data given in Table 2.1. Calculate Cronbach's α for the data in Table 2.2.

3.3   The following covariance matrix is for three successive measures of United States Party Identification (Markus, 1979). The corresponding correlations are given in parentheses. Calculate coefficients of reliability and stability based on the assumptions of the Wiley and Wiley (1970) model.

		1972	1974	1976
	1972	4.1290		
Year	1974	3.3395 (0.807)	4.1453	
	1976	3.2406 (0.890)	3.3663 (0.819)	4.0804

3.4   Use the variances and covariances provided in the above question (3.3) to calculate Cronbach's alpha for the three measurements.

3.5   Derive expected values for the variances and covariance of two τ-equivalent tests. Use these to derive expressions for the separate reliabilities of these two tests. Use these expressions to determine the reliabilities of the repeated measurements given in Table 1.1. Compare these with the simple Pearson correlation between the two time points. Calculate the reliability of the sum of the two sets of measurements in Table 1.1 under the assumption of either τ-equivalence or parallelism.

4

Designs for Reliability Studies

4.1 Introduction

The design of a reliability study is clearly dependent on the context within which the study is being undertaken, what sort of measuring instruments are being used or compared, and what properties or characteristics are being measured. There appear to be three main roles for reliability studies. These are (a) as an aid to instrument development (including training of interviewers, raters or examiners), (b) as an aid to the choice of measuring instrument or to the choice of conditions under which measurements are to be made and (c) as a way of monitoring instrument use (quality control). The proponents of generalisability theory distinguish generalisability studies (G-studies), which are used as the part of the development of a measuring instrument, from decision studies (D-studies) which involve the application of a measurement technique in a particular setting.

Consider the development and use of a structured psychiatric interview, a social questionnaire, or a psychometric test. Early versions are piloted on suitable groups of subjects in order to assess such areas as agreement between raters or coders, the internal consistency of the subjects' responses, their stability over time, and so on. Difficulties over the interpretation of instructions and the rating of responses are detected and ironed out. As a result of one or more of these pilot studies the measuring instrument is modified and improved. Eventually, the final form is assessed through the use of a formal reliability study. The properties of the instrument are published so that potential users can then decide on the suitability of the instrument for their own particular purposes and also make comparisons with competing measurement techniques. It is not enough, however, for a user to select a measurement technique and then simply proceed to use it in a routine way without evaluating its performance in its new setting. The user has to convince himself and others that the instrument is being used correctly and that it is performing as well as expected. The new user may need training and the instrument might have slightly different characteristics in its new role when compared to those observed by its developers. A particularly important aspect of its use is the population of subjects that it is to be used on. It is vital that the properties of a measurement technique are evaluated in every setting in which it is to be used and also that its behaviour is monitored during its routine use in those settings.

The development of a psychological or social measurement technique may appear to be particularly difficult to a physical scientist, but physical measurements need to be monitored and evaluated in an equally rigorous way. Analytical chemists, for example, routinely carry out *precision experiments* to evaluate sources of variation in measurements both within laboratories (*repeatability*) and between different laboratories (*reproducibility*). In the assessment of repeatability they are concerned with the closeness of results obtained on the same test material by the same technician using the same reagents, conditions and apparatus over reasonably short intervals of time. In the assessment of reproducibility, on the other hand, they are concerned with the influence of changing batches of reagents, conditions, technicians, apparatus, laboratories, and so on (Caulcott and Boddy, 1983). In the behavioural sciences the equivalent to a chemist's precision study is the generalisability study. Both types of study are concerned with potential sources of error and disagreement. Both types of study are explicitly concerned with the different facets of the measurement process and with the determination of the conditions under which measurements should be made.

Precision or generalisability studies may be used in both the development of a measurement method and the monitoring of its routine use. Returning to the area of instrument development, both physical and behavioural scientists are often interested in the comparison of multiple indicators of a particular property or trait or with the construction of a composite measurement of that property or trait. A well-known example from psychology is the comparison of scores obtained from several different tests of cognitive ability, all assumed to be indicators of a subject's intelligence. A familiar composite measure of intelligence is the Wechsler deviation intelligence quotient which is derived from scores obtained on eleven different indicators of cognitive ability (Wechsler, 1981). A typical example from analytical chemistry is the determination of iron in solution. This determination can be carried out gravimetrically, volumetrically, spectrophotometrically and in a number of other ways (Mandel, 1964). Perhaps a closer analogue of the measurement of intelligence is the use of various methods to measure the hardness or strength of materials. A more trivial example is that concerning the estimation of the lengths of pieces of string (Section 1.2).

4.2   Choice of subjects

When one thinks of a measuring instrument, and is concerned with the assessment of its potential usefulness, it is self-evident that one should consider the population of subjects or objects that it is to be used on. A particular weighing machine's usefulness will depend on whether it is to be used to weigh sacks of potatoes or small packets of spices, or whether it is to be used for weighing scrap iron or gold, and so on. Similarly a test of cognitive ability might be useful in discriminating between the talents of individuals in a general population of school children, but be practically useless when it is used in an attempt to monitor subtle changes in the abilities of a population of mentally handicapped children. The key question here is: 'Is the measuring instrument or technique capable of adequately fulfilling its intended role?' In order to answer this question an investigator is obliged to evaluate the method on subjects that are typical of those that will be encountered during its routine use. If a psychometric test is to be used for the comparison of gifted adults then its reliability should be assessed using a sample of these adults. If it is to be used for the assessment of children suffering from Down's syndrome then its reliability should be assessed from its use on a sample of children with Down's syndrome. A reliability study should be undertaken in the setting in which the instrument is to be used. It is not sufficient to rely on published details of an instrument's performance. There are often unforeseen ways in which an instrument's behaviour changes from one setting to another and the published details are unlikely to apply to the particular conditions and subjects in its new setting.

To emphasise the above point consider a further example. Here the investigator wishes to use the responses obtained during a structured psychiatric interview to allocate each subject or patient to one of several mutually exclusive diagnostic categories. The instrument (the psychiatric interview) is developed and evaluated using patients attending an out-patients' clinic of a large psychiatric hospital. Now, it should be obvious to the reader that, however good this interview is for use in a psychiatric clinic, if it is to be used in a completely different environment such as a general practice surgery or as part of a community survey, then it should be evaluated in this new environment prior to its routine use. The fact that there appears to be good agreement between clinicians working in a psychiatric clinic does not imply that there will be good agreement in general. Clinicians in the psychiatric clinic, for example, have the advantage that they can assume that the patient is suffering from some form of psychological distress. This is clearly not the case in general.

In selecting subjects for a study of agreement over diagnostic labels for psychiatric problems it should be borne in mind that the resulting agreement indices will be dependent on the range of subjects chosen. Agreement may be different for male and female subjects. It may depend on the subjects' age, ethnic background or social class. It will depend on whether the sample contains subjects without psychological problems or on whether the sample contains only patients with clear symptoms of psychiatric disturbance. The range of psychiatric disorders considered and their relative frequencies in the sample chosen for the study will also influence the outcome. It cannot be stressed too often that the value of a measurement instrument or a diagnostic technique, particularly its reliability, is not a unique characteristic of that instrument or technique. It is a function of the setting in which it used, the way it is used and of the population of subjects, objects or patients that are to be assessed.

4.3 Selection of measuring instruments and measurement conditions

In any given series of reliability or generalisability studies one might wish to compare the performance of measurements produced by instruments or techniques that are recognised as being qualitatively different (spectrophotometrical as opposed to volumetrical determination of iron in solution for example). These would be regarded by the layman as different measuring instruments. The use of a common type of spectrophotometrical method in several different laboratories in a precision study would, on the other hand, be recognised on a comparison of measurement conditions. Here different laboratories would not be regarded as different measuring instruments. The distinction between measuring instruments and measurement conditions is not, however, always as clear-cut as this. The subjects who estimated the lengths of the pieces of string (Table 1.2) might be considered to be three different measuring instruments. Similarly, a group of pathologists who independently assess the characteristics of biopsy material might also be regarded as a sample of different measuring instruments. If these pathologists were all to use the same assessment procedures to yield their conclusions, they might equally sensibly be regarded as examples of different measurement conditions. In some situations it might not be useful to use either of these terms. A group of psychiatrists independently rating a video recording of a structured psychiatric interview would not be regarded as representing different measuring instruments nor would they be considered as corresponding to different conditions of measurement.

Following the introduction to generalisability theory in Section 3.6 a particular sample of observations or measurements will be regarded as a sample from a universe of admissible observations, characterised by one or more *facets*. A facet can include alternative measuring instruments or test forms as implied above, or it can cover different measurement conditions such as home, school or psychiatric clinic, time of day, and so on. Other facets of observation include the identity of the rater, clinician or technician and type of information being used to make a rating (a brief written vignette, a video-recording or direct observation of a psychiatric interview, for example). Each facet of observation will have two or more possible *levels*. These levels can define different measuring instruments, different measurement conditions, different interviewers and/or raters, and so on. An important component of the design of a reliability, generalisability or precision study is a decision concerning the number of facets of the observations that will be systematically varied and also the number of levels for each of the facets chosen. In Section 3.6 a simple example of a two-facet design was given. Here three raters each assessed subjects' anxiety in the morning and again in the evening. The first facet (rater) was chosen to have three levels and the second (time of day) was chosen to have two.

In Chapters 5–7 the analyses to be described will concentrate on reliability studies with one of two major aims. The first is the comparison of distinct measuring instruments or alternative forms of a particular type of instrument where the goal is to assess the consistency with which the instruments jointly measure a common trait, the relative biases of these instruments, and also their precisions. The second is the assessment of inter-rater agreement or disagreement either through the use of variance components models (interval data) or indices of agreement (nominal, ordinal or interval data). In the latter case a 'rater' might be an alternative form of a questionnaire or test so that the distinction between the two aims is not always clear-cut.

An important consideration in the interpretation and subsequent use of the results of a reliability study is the realization that the levels of a given facet of measurement can be thought of as either being fixed or as a random selection of a potentially infinite universe of levels. In the former case one is interested only in the comparison of instruments, conditions or raters that have been specified in the reliability study. In the latter case one might be interested in agreement between raters or measurement methods in general, not just the ones that are used in the particular reliability study at hand.

To clarify the last point, consider the design of a reliability study to assess diagnostic agreement between psychiatrists. In the main research or decision study it is intended that each subject or patient will be interviewed and given a diagnostic label by one of a panel of four research psychiatrists. In the preliminary reliability or generalisability study each subject or patient is given a psychiatric diagnosis by each of the four psychiatrists. Here the interest is only in the relative performance of these four assessors. The facet (psychiatrist) can be considered to have four levels that are fixed. Now let the design of the decision study change. In the new design it is recognised that each patient might be assessed by one of a panel of, say, up to 200 psychiatrists. In the preliminary reliability study a sample of ten psychiatrists takes part. Here the primary interest of the reliability study is not in the performance of the ten psychiatrists who have actually taken part, but in the performance of psychiatrists in general. In this reliability study the ten levels of the facet (psychiatrist) are not fixed but random.

The above discussion has implications for the design of a reliability or generalisability study. If the aim is simply to compare two or three measuring instruments or fixed raters then one simply uses these two or three instruments or raters to characterise a sample of subjects. These instruments or raters may, of course, be used under different conditions (that is, there may be other facets of the observations to be investigated in the reliability study). If one wishes to draw inferences concerning the agreements amongst raters in general, however, one usually aims to have considerably more than two or three raters take part in the reliability study.

4.4 Simple replications

The simplest design for a study of instrument precision involves repeated measurements on the same subjects using the same instrument under conditions that remain as stable as possible. Examples using a pair of kitchen scales were described in Sections 1.1 and 1.2. Perhaps a more realistic estimation of the variability of measurements would be obtained by allowing the measurement conditions to vary in an unsystematic way. Resetting the zero adjustment was a source of variation explicitly discussed in Chapter 1. If measurements are repeated with the same zero adjustment or they are made using the same batch of chemical reagents, then the measurements will show relatively low variability. The estimate of precision might be too optimistic. Jaech (1979) discusses these problems in the context of the assessment of within-laboratory variability from data provided by a precision study. Individuals who have experience in making measurements realise that there are several sources of error within a laboratory. Under the assumption that the quality of a laboratory's measurements will be assessed from an estimate of repeatability (see Section 4.1) the analysts will try to hold constant as many of the sources of error as possible. Although the resulting replicate data may provide a valid estimate of repeatability (as defined in Section 4.1), this estimate of repeatability will inevitably be an underestimate of the within-laboratory variability. The latter should reflect all sources of random error that are usually associated with a repeated measurement.

It is usual to assume in these simple precision studies that errors of measurement are statistically independent. If the measuring instruments are human raters (psychiatrists diagnosing different forms of madness in a group of psychotic patients, for example) then it is quite likely that the raters will remember their previous ratings or diagnoses and their assessments will be influenced by them. Usually they will give similar ratings or diagnoses in each of the replications. The results would again suggest that raters' performance is better than it really is. This effect of memory is not limited to the case where raters make subjective decisions. In a second attempt at a test of cognitive ability, a subject may remember his or her previous answers and simply repeat them. Someone filling in a questionnaire on political attitudes for the second time is also likely to remember the responses from the first time around and repeat them to give a potentially misleading impression of stability.

Finally, as soon as one begins to make measurements on living organisms, it is clear that the characteristics being measured might also change from one replication to another. In the behavioural and social sciences one needs to take measurements fairly close together (in time) so that there is little possibility of any underlying change, but one is also conscious of the fact that if they are too close the subject (and the rater) may simply remember the previous response or rating and repeat it. In Section 3.4 a way of disentangling instability and lack of precision was discussed but, again, this was dependent on the assumption that measurement errors were uncorrelated. More complex analyses would be required to cope with correlated measurement errors (see Wertz *et al.*, 1980).

4.5 Multiple indicators

In very many practical situations it is simply impossible to measure a characteristic at more than one time. It should be clear, for example, that if one requires a measurement of a reaction time, then the reaction time cannot be generated again for further measurement. The only way out of this problem is to make simultaneous measurements using alternative measuring devices (multiple indicators). Interesting examples are provided by Grubbs (1948, 1973) who is concerned with such problems as the estimation of the velocities of field-gun shells or with fuse-burning times. Clearly for one particular shell or explosive device there is only one opportunity to make the required measurements!

Returning to the life sciences and, in particular, to the behavioural and social sciences, it is clear that many characteristics are not stable. One might, for example, wish to measure anxiety at a given point in an experiment or immediately after presenting a subject with an aversive stimulus. Typically, a psychologist would wish to measure anxiety just before presentation of a spider to a spider-phobic and again just afterwards. In this situation anxiety is a transient characteristic of a particular experimental condition (state anxiety as opposed to trait anxiety) and, again, it needs to be assessed simultaneously through the use of two or more indicators. These indicators could be behavioural (how close the subject is prepared to go to the source of the stimulus, or how anxious he or she claims to be) or physiological (heart rate, breathing rate, or skin conductivity). Other transient characteristics that might require the use of simultaneous multiple indicating are fatigue, depression, hunger, thirst and arousal.

Multiple indicators might be required for a further reason. In Section 1.3 factor analytic models were introduced as a way of defining the concept being measured. The measurement model was inferred from the correlations between scores on tests of cognitive ability, for example, and the idea of general intelligence arose from the concept of a latent trait or factor common to these cognitive tests (Spearman, 1904). Here, general intelligence itself is not directly measureable. Similarly, it is impossible to obtain direct measurements of neuroticism, ambition, social status or racial prejudice. Again, however, these concepts can be measured indirectly through the use of multiple indicator variables. Racial prejudice, for example, might be assessed from responses to several questions such as 'Do you have any friends who belong to a different ethnic background to your own?', or 'Do you agree with legislation to prevent racial discrimination?', and so on. The use of multiple indicators in this way is not, of course, restricted to the behavioural or social sciences. Examples from the physical sciences include different methods of measuring iron in solution or the methods of measuring hardness or strength of materials (see Section 4.1).

4.6 Randomised blocks designs

Consider an investigation of inter-rater agreement in which each of I subjects is independently rated by each of K raters, diagnosticians or examiners. The investigation yields a total of IK measurements or ratings. Here ratings on a single subject provided by each of the K raters are regarded as a profile of scores provided by a simple set of multiple indicators. A simple example from medicine might involve the use of five pathologists (raters) to independently grade each of a hundred histological specimens (subjects). Similarly, several radiologists might be required to characterise each of a sample of lung X-ray images. Tables 2.1 and 2.2 summarize the independent assessments of 29 candidates' scripts by just two examiners. The string measurements on Table 1.2 provide another simple example.

Suppose that each subject is to be independently assessed by each of five raters, or equivalently, each subject is to be asked to attempt five different tests of cognitive ability. It may be possible for the five raters to simultaneously assess each subject but it will be assumed for the time being that this is not the case (each assessment, for example, might involve each rater interviewing each subject). Clearly subjects could not be expected to simultaneously attempt five different ability tests! The order in which the ratings are carried out or in which the tests completed might have an important influence on the results and it is therefore good practice for the order to be randomised. The resulting design is a *randomised blocks design* (Cochran and Cox, 1957; Cox, 1958) where the subject is equivalent to a block, the time of rating or testing is equivalent to a plot, and the rater or test form is comparable to an experimental condition or treatment.

Psychiatry is one area of medicine where rater agreement studies are regarded as being particularly important. Here a sample of clinicians (raters) can each assess the patients (subjects) using one of several alternative rating procedures. Each clinician for example might independently rate a written case vignette or interview transcript from each of the sample of subjects. Here the different clinicians would simply be rating or scoring information from an interview carried out at a different time and perhaps by a clinician who is not part of the study itself. If this procedure were thought to be unsatisfactory then another approach might be for the clinicians to observe video-taped interviews or even watch the interview as it takes place (either from within the interview room or through a half-silvered mirror).

Now suppose a study design is chosen in which psychiatrists are intended to sit-in on interviews directed by another clinician. Each psychiatrist simply watches and listens to the interview as a neutral spectator and then independently makes an assessment of the patient's psychiatric condition. This design is satisfactory from many points of view but there is often an important practical constraint imposed which makes it unrealistic. Suppose that the interview room is too small to hold more than three spectators or that the patient and interviewer are incapable of tolerating the presence of more than two spectators. Alternatively, suppose that it is far too expensive to plan for all psychiatrists to attend all interviews.

The way to cope with the above constraint is through the use of a *balanced incomplete blocks design (BIBD)*. This is a design originally proposed by Yates (1936) and is described in terms of its use in reliability studies by Fleiss (1981a, 1986). An example of its use in the development of a social maladjustment and dysfunction schedule is provided by Clare and Cairns (1978). Another example of where this type of constraint will be imposed is where each of the clinicians (whether in psychiatry or another branch of medicine) is expected to interview or examine each patient independently. Here the patient might not tolerate more than two examinations (it will be assumed that it is possible for the patient to have two independent examinations).

The use of BIBD in reliability studies will be illustrated using the data given in Table 4.1 (from Fleiss, 1981a). Here each of ten subjects is rated for depression by three raters. The study involves a total of six raters and each of these raters rates five subjects. Finally, it can be seen that each pair of raters jointly rates two subjects. In the study reported by Clare and Cairns (1978) each of

Table 4.1 Results of a BIBD reliability study of a rating scale for depression*

	Subject									
Rater	1	2	3	4	5	6	7	8	9	10
1	10	3	7	3	20					
2	14	3				20	5	14		
3	10		12			14			12	18
4		1		8			8		17	19
5				5	26	20		18	12	
6		9			20		14	15		13

* Reproduced with permission from Fleiss (1981a).

48 subjects were assessed by two raters. The study involved the use of a total of four raters and each of these raters assessed 24 subjects. Each distinct pair of raters jointly rated eight subjects.

Let, in general, K denote the number of raters or examiners involved in the reliability study, I the total number of subjects being rated, P the number of raters rating any subject ($P < K$), Q the number of subjects rated by any one rater ($Q < I$), and R the number of subjects rated by any pair of raters. Conditions that must be fulfilled for any reliability design to be a BIBD are the following.

$$KQ = IP \qquad (4.1)$$

$$K \leq I \qquad (4.2)$$

$$R(K - 1) = Q(P - 1) \qquad (4.3)$$

For the design described by Fleiss (1981a) $K = 6$, $I = 10$, $P = 3$, $Q = 5$ and $R = 2$. For that described by Clare and Cairns (1978) $K = 4$, $I = 48$, $P = 2$, $Q = 24$, $R = 8$. Fleiss (1986, p. 293) provides a table of balanced incomplete blank designs for reliability studies involving up to six raters. Cochran and Cox (1957, pp. 469–82) present plans for balanced incomplete blocks designs which can be used in reliability studies involving up to 28 raters.

The BIBD is particularly useful when one is interested in looking at relative biases of raters through comparisons of their means and when each rater's mean is required to be estimated with the same precision. There is one serious drawback to the use of a BIBD, however, and that is the possibility that one or more of the raters might fail to make the planned assessments. If the primary interest of the reliability study is in estimation of reliability coefficients (e.g. intra-class correlations) rather than in comparisons of rater's means, or if there is a possibility of missing ratings, then an alternative design should be considered. This design simply involves the selection of a simple random sample of P raters (for the total of K) separately and independently for each of the subjects (Shrout and Fleiss, 1979).

Returning to the randomized block design with no missing data, there is one other potential complication which might influence the results. In the description of the randomised blocks design given above it was assumed that in the case of assessments based on interviews the interviewer was not providing ratings for the study. Raters were simply assumed to be neutral observers of either live interviews or recordings of interviews. In practice, however, this might not be the case. Typically if K raters assess each subject one of these K raters will conduct the interview *and* provide a rating, and the other ($K - 1$) raters will provide ratings only. If the design is balanced, then each of the K raters interviews a separate sample of I/K subjects, where I is the total number of subjects taking part in the reliability study. In the study of Clare and Cairns (1978), for example, each subject received two ratings on each item of the social maladjustment schedule, one from an interviewer and one from the accompanying rater who simply observed and rated. Fleiss (1970) gives an example of a study using three interviewers/raters where each subject is assessed by an interviewer and two independent observers. This form of design is an example of a *partially nested design* with interviewers crossed with raters and subjects crossed with raters but nested within interviewers. Crossing and nesting will be dealt with in detail in Section 4.7. Here it is only necessary to point out that a subject's rating can be dependent on the identity of the interviewer (interviewer effect) and of the rater (rater effect). There is also a possibility of various interviewer–rater interaction effects. The reader is referred to Fleiss (1970) for details. Clare and Cairns (1978), like most investigators, chose to ignore possible complications caused by simultaneous interviewing and rating.

4.7   Crossing and nesting

The discussion in this section will begin by returning to a few simple examples of precision studies in the physical sciences. Caulcott and Boddy (1983, Chapter 9) give an example of a precision study involving measurement of ammonia in solution. Four batches of ammonia solution are prepared (at concentrations of 10%, 15%, 20% and 25% by mass). Four laboratories are selected at random from a population of 30 laboratories and each of these 4 laboratories is sent 5 samples at each of the 4 concentrations. This study furnishes 80 measurements (20 from each of the 4 laboratories). This is an example of a fully *crossed* design. Each laboratory provides measurements on all four batches (concentrations) of ammonia solution. Mandel (1964) provides a description of a similar design for the measurement of the smoothness of 14 batches of paper by 14 different laboratories.

Now consider a superficially similar experiment where three laboratories are each sent six identical samples for assay. Each laboratory routinely makes measurements in large batches (corresponding to a particular preparation of test reagents, for example) and in the precision study each laboratory is asked to insert two of the test samples into each of three batches of measurements (Caulcott and Boddy, 1983, Chapter 10). This design is an example of a simple *nested* or *hierarchical* design. Batches are not crossed with laboratories in this case; batch 1 of laboratory A, for example, is not the same as batch 1 of laboratory B. There are, in fact nine different batches, not three. Mandel (1964) provides an example of a hierarchical design in which two operators from each of three laboratories provide duplicate measurements on each of three different days. Replicates are nested within days and days are nested within operators who are, in turn, nested within laboratories.

Simple examples of crossed designs in the social and behavioural sciences are proved by repeated testing using the same multiple indicators. A test–retest reliability study of cognitive performance using a fixed battery of psychometric test forms would provide data that are cross-classified by period of testing and test form. This would only hold, however, if the first testing of each of the subjects occurred at roughly the same time and the same should hold for the second round of testing. This test–retest design, of course, is equivalent to two-wave multiple indicator panel study of the social sciences (Wheaton *et al.*, 1977). Three- and four-wave multiple indicator panel studies are also examples of this type of design.

A study of the reliability of measures of psychological impairment in disaster survivors illustrates the use of a fully crossed design (Gleser *et al.*, 1978). Twenty adult survivors were each interviewed independently by two interviewers. Two raters then independently scored the interview data by rating each survivor on a number of subscales (such as anxiety, for example). Taking scores for anxiety this crossed design yields 80 measurements (20 subjects $\times$ 2 interviewers $\times$ 2 raters).

Brennan (1975) has discussed the use of nested designs to ask the question, 'How should one estimate the reliability of school (or classroom) means when persons are nested within schools (or classrooms)?' This is clearly analogous to the use of batches nested within laboratories in the analytical chemists' precision experiments (q.v.). A further example could arise from comparison of inter-rater agreement within one clinical team (Clinic A) with that within another (Clinic B). Assume that there are 20 patients to be assessed by each of 5 clinicians in Clinic A and a further 20 patients to be assessed by each of 5 clinicians in Clinic B. Patients are crossed by clinicians but both patients and clinicians are nested within clinics. This design yields 200 observations (20 $\times$ 5 plus 20 $\times$ 5). If the behaviour or symptoms of the patients were recorded in written summaries or as video-taped interviews there would be no reason why all 10 raters should not each assess all 40 patients (yielding 400 observations). Patients and clinicians are now fully crossed but clinicians are still nested within clinics. Cronbach *et al.* (1972) give a further example from education where teachers are rated by a team of judges, but a different set of judges is used for each school. That is, judges are nested within schools and so are school teachers. The reader is also referred to the monograph of Cronbach *et al.* (1972) for detailed discussions of more complicated designs which are beyond the scope of the present text.

4.8 'Case-control' sampling

The construction of simple calibration curves using a series of known concentrations of a drug or other chemical compound was briefly discussed in Section 1.7, which also contained the derivations of the specificity and sensitivity of a psychiatric screening questionnaire such as the GHQ (Goldberg, 1972). Here the results of scoring the GHQ responses are compared with the supposedly infallible assessment provided by a detailed psychiatric interview. One could, of course, acknowledge that both instruments are fallible and proceed to calculate a measure of agreement such as Cohen's κ (Section 2.8).

The GHQ is simple and cheap to administer. A structured psychiatric interview, on the other hand, is time-consuming and therefore expensive. For this reason the 'validation' (calibration) of the GHQ is usually carried out using a two-stage sampling scheme. First, a large sample of subjects is tested using the GHQ. Subjects are labelled as 'GHQ + ve' and 'GHQ $-$ ve' as a result of their responses to this screening questionnaire. A sample of the GHQ + ves and an independent sample of the GHQ $-$ ves are then given a detailed psychiatric interview to determine whether or not the subjects are really suffering from psychiatric problems. Usually all of the GHQ + ves and about half of the GHQ $-$ ves are interviewed (Tarnopolsky *et al.*, 1979). The properties of the GHQ are then determined from the resulting 2 $\times$ 2 summary table.

Jannarone *et al.* (1987) have investigated the above 'case-control' sampling scheme in considerable detail. They describe the sampling scheme in the following way.

Instead of a random selection of N individuals, a large number is sequentially sampled until at least $N/2$ have been classified as positive and $N/2$ have been classified as negative by rater 1. (To the extent that rater 1 classification probabilities differ from .5, larger samples will be necessary before obtaining at least $N/2$ in each cell). Next $N/2$ individuals are randomly selected from each of the resulting positive and negative rater 1 classification groups for classification by rater 2.

Here rater 1 is equivalent to the GHQ and rater 2 is equivalent to the structured psychiatric interview. From the resulting data one can estimate the *positive predictive value* (A) of the GHQ from the proportion of GHQ + ves that are found to be cases from the results of the structured psychiatric interview. Similarly, one can estimate the *negative predictive value* of the GHQ ($1 - C$) from the proportion of the GHQ $-$ ves found to be non-cases from the results of the interview. The proportion C is the complement of the negative predictive value. If p is the probability that an individual drawn at random from the appropriate population will be found to be GHQ + ve (with $q = 1 - p$), then κ can be calculated from

$$\kappa = \frac{2pq(A - C)}{p + (q - p)(Ap + Cq)} \tag{4.4}$$

Jannarone *et al.* (1987) use the δ technique (see Appendix 3) to provide variances for the estimates of κ based on estimates of A and C in the cases when p is estimated from the data (and also when it is not estimated from the data).

These authors then proceed to estimate the sample size requirements for a test of whether κ is significantly different from zero (with a significance level set at 0.05 and a power of 0.95). Their results are shown in Table 4.2. The sample sizes required are always less for the case–control sampling plan than for the cross-sectional sampling method. For values of p greater than 0.90 and for values of κ less than 0.70 the performance of the case–control sampling design is much better. Note, however, that these conclusions apply only if the cost of the first rating (the GHQ for example) is negligible when

Table 4.2 Required sample sizes for two approaches to classification agreement assessment[*]

True kappa	Cross-sectional sample size	Case–control sample size $p^†$				
		0.5	0.7	0.9	0.95	0.99
0.10	1300	1296	1124	606	416	236
0.30	144	140	128	90	74	56
0.50	52	48	46	36	32	30
0.70	24	22	22	20	18	18
0.90	13	12	12	10	10	10

[*] Reproduced with permission from Jannarone *et al.* (1987). Entries are required second rating-method sample sizes for power levels of 0.95, given a two-tailed, 0.05 level of test association. Case–control entries are rounded to nearest even integer and cross-sectional entries are rounded to the nearest integer.
[†] p is the common marginal probability that either rating method will yield a positive rating.

compared to the second. Table 4.2 shows, for example, that for $p = 0.9$ and $\kappa = 0.5$ only 36 second ratings are required for the case–control sampling method, compared with 52 from the cross-sectional sampling design. However, the first rater would have had to look at 360 subjects in order to find the 36 suitable subjects for the second rater.

4.9 Sample size requirements

Table 4.2 is the result of one of a very few studies of the statistical power of reliability studies. The most detailed appears to be that of Donner and Eliasziw (1987). These authors consider sample size requirements in simple replication studies involving K raters and I subjects. The measure of reliability used is the intra-class correlation estimated through the use of a one-way analysis of variance (see Chapter 6). Other more complex designs would require different sample sizes, but the results of Donner and Eliasziw will provide a general impression of how the behaviour of an agreement statistic changes with increases on numbers of subjects (I) and/or ratings (K). More detail for the more complex designs will be given in the later chapters. Kraemer and Korner (1976) have considered sample size requirements for test–retest data, and Kraemer (1976) has also investigated this problem when the number of subjects (I) *or* the number of raters (K) is large. The study of Donner and Eliasziw deals with power requirements as both I *and* K vary.

Donner and Eliasziw based their power calculations on the assumption that interest focuses on the null hypothesis that the intra-class correlation, r_i, is less than or equal to some value p versus an alternative hypothesis that r_i is greater than p. These authors generated contours of equal power for a range of null and alternative hypotheses as a function of the true value of r_i. Figure 4.1 is an example of one of their sets of power contours.

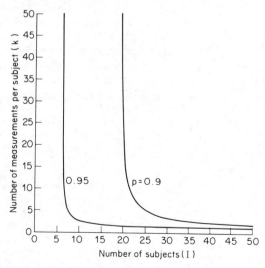

Fig. 4.1 Power contours for testing null hypothesis that $r_i = 0.8$ versus the alternative that $r_i > 0.8$ using a significance level of 0.05 and power of 0.80 (reproduced with permission from Donner and Eliasziw (1987)).

Figure 4.1 shows two power contours – one for a true r_i of 0.90 and the other for a true value of 0.95. In both cases the contours were generated using a significance level of 0.05 and a required power of 0.80. The null hypothesis was that $r_i = 0.80$ versus an alternative hypothesis that r_i was greater than 0.80. Technical details of how these contours were generated are given in Donner and Eliasziw (1987).

Figure 4.1 illustrates the relative influence of I and K on the achieved power. The first point to note is the asymptotic nature of the curve as I increases. When $r_i = 0.8$, for example, there is barely any increase in power in a test–retest study (i.e. $K = 2$) as the sample size (I) is increased above 15. The results also demonstrate that the required value of K for a given I increases very rapidly as I is lowered. The actual choice of I and K will, of course, depend on the relative problems concerning recruitment of subjects and the ability to provide several independent replicates of the measurements. In general, however, Donner and Eliasziw conclude that an increase in K for fixed I provides more information than an increase in I for fixed K.

5

Covariance-Components Models

5.1 Introduction: a model for string lengths

Reconsider the estimates of the 15 pieces of string given in Table 1.2. Assume, for the purposes of this discussion, that we do not have access to the 'true' measurements; that is, we only have Graham, David and Brian's estimates of the lengths of the pieces of string. The first step in the analysis of data of this sort is to calculate simple summary statistics and plot graphs to visually examine the relationships between the three estimates. Pairwise scatter diagrams will give some indication of how the estimates are related, how variability on the estimates does or does not appear to change with the perceived lengths of the pieces of string, and so on. Another very important role of the preliminary graphical displays is to check whether there are any outliers which may affect any subsequent statistical analyses. At this stage the investigator might choose to discard any outlying observations after first, of course, considering possible reasons for their existence. Alternatively, the outliers might be retained for the time being and any subsequent statistical analyses carried out with or without these outliers contributing to any summary statistics used in these analyses. Plots for the data in Table 1.2 (not shown) confirm the expected linear relationships between the three estimates and also that there are no obvious outliers.

Summary statistics for the data in Table 1.2 are given in Table 5.1. It can be seen that these three estimates are very highly correlated. There is also an increase in the variance of the estimates going from Graham to Brian, and from Brian to David. There is a hint that this might also be reflected in a parallel change in the means of the estimates.

In Chapter 1 a model was postulated to explain the interrelationships between the three sets of estimates. This had the form

$$X_{ik} = \alpha_k + \beta_k \mu_i + e_{ik} \tag{5.1}$$

where X_{ik} is the kth estimate of the length of the ith piece of string ($k = 1$ for Graham, 2 for Brian and 3 for David, $i = 1$ to 15). Here μ_i is the unknown 'true' length of the ith piece of string and e_{ik} is the error at measurement made by the kth estimate for this piece of string. In this model X_{ik}, μ_i and e_{ik} are all assumed to be random variables. The parameters of the model, that is, α_k

Table 5.1 Summary statistics for estimates of string lengths (Table 1.2)

(a) Covariance matrix

	Graham	Brian	David
Graham	2.121		
Brian	2.267	2.655	
David	2.569	2.838	3.235

(b) Correlations

	Graham	Brian	David
Graham	1		
Brian	0.9553	1	
David	0.9807	0.9684	1

(c) Means

	Graham	Brian	David
	3.433	3.427	3.767

and β_k jointly describe the measurement bias characteristic of subject k. α_k and β_k are equivalent to the intercept and slope respectively, of a straight line relating the kth subject's length estimates with the latent true scores. The final parameter of the model is $\bar{\mu} = E(\mu_i)$. If it is further assumed that both $\text{Cov}(e_{ik}, e_{jm}) = 0$ for any i, j, k or m (except when $i = j$ *and* $k = m$) and $\text{Cov}(e_{ik}, \mu_j) = 0$ for any i, j or k, then the expected value of variance of X_{ik} is given by Equation (1.20), that is

$$\text{Var}(X_{ik}) = \beta_k^2 \sigma^2 + \sigma_k^2 \tag{5.2}$$

where σ^2 and σ_k^2 are model parameters representing the variance of μ_i and of e_{ik}, respectively. Similarly, the expected value of the covariance of X_{ik} and X_{im} is given by Equation (1.21), that is

$$\text{Cov}(X_{ik}, X_{im}) = \beta_k \beta_m \sigma^2 \tag{5.3}$$

The expected dispersion or covariance matrix for the estimates of string lengths given in Table 1.2 is therefore of the form given in Table 5.2. To summarise, each variance and covariance can be expressed as a function of the model parameters β_1, β_2, β_3, σ_1^2, σ_2^2, σ_3^2 and σ^2. The entries of the covariance matrix can be partitioned into different components and hence the term *covariance-components model*.

Ignoring the intercept terms (the α_k) for the time being, one can approach the problem of estimating the parameters of the model in Equation (5.1) by equating the observed and expected values of the entry of the covariance

Table 5.2 Expected covariance matrix for the estimates of string length (Table 1.2)

	Graham	Brian	David
Graham	$\beta_1^2 \sigma^2 + \sigma_1^2$		
Brian	$\beta_2 \beta_1 \sigma^2$	$\beta_2^2 \sigma^2 + \sigma_2^2$	
David	$\beta_3 \beta_1 \sigma^2$	$\beta_3 \beta_2 \sigma^2$	$\beta_3^2 \sigma^2 + \sigma_3^2$

matrix from the three estimates and solving the system of six independent simultaneous equations provided by the three variances and the three unique covariances. There is immediately a problem in this approach, however. The expected covariance matrix as described so far has seven parameters and therefore no unique solutions can be obtained. Representing the same covariance between estimate i and estimate j as S_{ij}, we have

$$S_{11} = \beta_1^2 \sigma^2 + \sigma_1^2$$
$$S_{22} = \beta_2^2 \sigma^2 + \sigma_2^2$$
$$S_{33} = \beta_3^2 \sigma^2 + \sigma_3^2$$
$$S_{12} = \beta_1 \beta_2 \sigma^2 \qquad (5.4)$$
$$S_{13} = \beta_1 \beta_3 \sigma^2$$
$$S_{23} = \beta_2 \beta_3 \sigma^2$$

These six equations can only be solved if we introduce a constraint on the parameter values. This can be done by fixing the value of one of the slope parameters ($\beta_1 = 1$, for example) or by fixing the value of σ^2 ($\sigma^2 = 1$, for example). If this is done there will then be a possibility of obtaining a unique estimate for each of the remaining free parameters.

Non-statisticians might feel a little uneasy about the introduction of an 'arbitrary' constraint such as those described in the above paragraph. Returning to the equations in (5.4) we can see that the ratio of β_1 to β_2 (that is β_1/β_2) can be estimated by S_{13}/S_{23} and similarly, $\beta_1/\beta_3 = S_{13}/S_{23}$ and $\beta_2/\beta_3 = S_{12}/S_{13}$. We can only describe the biases in Graham's estimates *relative to the estimates* of either those of Brian or David. Alternatively, we could describe the biases in Brian's or David's estimates relative to those of Graham. The ratios of the form β_k/β_m (for $k \neq m$) can be estimated but the separate β_k cannot. This is the problem of *identifiability* and the model without the constraints is referred to as being *under-identified*. If, on the other hand, we were to recognise that the μ_i are hidden or latent values with a completely arbitrary scale of measurement (remember that we do not have access to the 'true' values given in Table 1.2) then we could fix this scale of measurement by letting $\sigma^2 = 1$, for example. If the scale of measurement for μ_i is fixed in this way then unique estimates of β_k can be obtained.

Consider, first, the constraint $\beta_1 = 1$. The required estimates of the other six parameters are then obtained as follows:

$$\hat{\beta}_2 = S_{23}/S_{13} = 1.105$$
$$\hat{\beta}_3 = S_{23}/S_{12} = 1.252$$
$$\hat{\sigma}^2 = \frac{S_{12} S_{13}}{S_{23}} = 2.052 \qquad (5.5)$$
$$\hat{\sigma}_1^2 = S_{11} - \hat{\sigma}^2 = 0.069$$
$$\hat{\sigma}_2^2 = S_{22} - \hat{\beta}_2^2 \hat{\sigma}^2 = 0.151$$
$$\hat{\sigma}_3^2 = S_{33} - \hat{\beta}_3^2 \hat{\sigma}^2 = 0.019$$

If, instead of setting $\beta_1 = 1$, we constrain σ^2 to be fixed ($\sigma^2 = 1$), then the remaining estimates are given by

$$\hat{\beta}_1 = \sqrt{\left(\frac{S_{12}S_{13}}{S_{23}}\right)} = 1.433$$

$$\hat{\beta}_2 = \sqrt{\left(\frac{S_{12}S_{23}}{S_{13}}\right)} = 1.583$$

$$\hat{\beta}_3 = \sqrt{\left(\frac{S_{13}S_{23}}{S_{13}}\right)} = 1.783 \tag{5.6}$$

$$\hat{\sigma}_1^2 = 0.069$$

$$\hat{\sigma}_2^2 = 0.151$$

$$\hat{\sigma}_3^2 = 0.019$$

Note that, subject to rounding errors for both sets of constraints, $\beta_1^2\sigma^2 = 2.052$, $\beta_2^2\sigma^2 = 2.504$, $\beta_3^2\sigma^2 = 3.216$, $\beta_2/\beta_1 = 1.105$, $\beta_3/\beta_1 = 1.252$ and $\beta_3/\beta_2 = 1.133$. If we now consider the expected values of the X_{ik} we can attempt to estimate α_1, α_2 and α_3 together with $\bar{\mu} = E(\mu_i)$ by equating the three expected values with the corresponding sample means. Again, however, there is a problem of identifiability. Another constraint is needed. Either one of the α_k must be set to 0, or $\bar{\mu} = 0$. We have

$$\bar{X}_1 = \alpha_1 + \beta_1\bar{\mu}$$

$$\bar{X}_2 = \alpha_2 + \beta_2\bar{\mu} \tag{5.7}$$

$$\bar{X}_3 = \alpha_3 + \beta_3\bar{\mu}$$

If the constraints imposed on the model are $\beta_1 = 1$ *and* $\alpha_1 = 0$, then

$$\hat{\bar{\mu}} = \bar{X}_1 = 3.433$$

$$\hat{\alpha}_2 = \bar{X}_2 - \hat{\beta}_2\hat{\bar{\mu}} = -0.367 \tag{5.8}$$

$$\hat{\alpha}_3 = \bar{X}_3 - \hat{\beta}_3\hat{\bar{\mu}} = -0.531$$

However, if the constraints are $E(\mu_i) = \bar{\mu} = 0$ *and* $\sigma^2 = 1$, then

$$\hat{\alpha}_1 = 3.433$$

$$\hat{\alpha}_2 = 3.427 \tag{5.9}$$

$$\hat{\alpha}_3 = 3.767$$

What about the reliabilities of the three subjects' estimates? Whichever choice of constraint is used (that is, either $\beta_1 = 1$ or $\sigma^2 = 1$) the reliabilities can be estimated using Equation (3.7), that is

$$R_k = \frac{\text{Var}(\tau_i)}{\text{Var}(X_{ik})} \tag{5.10}$$

$$= \frac{\text{Var}(\beta_k\mu_i)}{\text{Var}(\beta_k\mu_i + e_{ik})} \tag{5.11}$$

$$= \frac{\beta_k^2\sigma^2}{\beta_k^2\sigma^2 + \sigma_k^2} \tag{5.12}$$

The required estimates for R_1, R_2 and R_3 are 0.968, 0.943 and 0.994.

5.2 Precision of reliability estimates

Having estimated the parameters of a measurement model such as that described by Equation (5.1), the investigator's attention should then turn to the problems of sampling variability of the estimates. The discussion could be relevant to any of the parameter estimates (or functions of them) obtained in the previous section but, for illustrative purposes, it will be restricted to a study of the sampling variability of the reliability of Graham's estimates (that is, to R_1). Clearly, if we were to repeat the estimates of string length for a different sample of pieces of string it is very unlikely that we would get exactly the same estimate of R_1 again. The estimate of R_1 is itself a random variable with a characteristic probability distribution. The aim of the present section is to describe ways in which the probability distributions of an estimate such as R_1 can be investigated without taking further samples or by recourse to difficult mathematics. In some cases the theoretical sampling distribution of a reliability or agreement coefficient will not actually be known and, in these situations, the method described below may be the only way in which the investigator can get a picture of the sampling variability of the estimate. A further advantage of the method to be described below is that it does not depend on detailed assumptions concerning the distributional form of the raw measurements, X_{ik}.

Let the vector $\mathbf{x}_i$ denote the three measurements X_{i1}, X_{i2} and X_{i3}. This vector has a characteristic, but usually unknown trivariate probability distribution. Furthermore, the probability distribution of a statistic such as R_1 is determined by this probability distribution of $\mathbf{x}_i$. Efron (1979) introduced a very general resampling procedure for investigating the sampling distributions of statistics based on the observed sample of observations $\{\mathbf{x}_i\}$. This method is called the *bootstrap* (see Diaconis and Efron, 1983; Efron and Gong, 1983; or Efron and Tibshirani, 1986). The procedure involves approximating the sampling distribution of a function of the observations (say, R_1) by what Efron calls the bootstrap distribution of the quantity. The bootstrap distribution is obtained by replacing the unknown distribution of $\mathbf{x}_i$ by observed or *empirical* distribution of the sampled measurements and then resampling many times (with replacement) from the observed distribution to obtain the estimated probability distribution of the required statistic.

The empirical distribution of the sample of string measurements in Table 1.2 is the distribution function of $\mathbf{x}_1, \mathbf{x}_2, \ldots, \mathbf{x}_{15}$ putting mass $\frac{1}{15}$ on each of the $\mathbf{x}_i$. A *bootstrap sample* of size 15 can be obtained by sampling *with replacement* from Table 1.2. Let $\mathbf{x}_1^*, \mathbf{x}_2^*, \ldots, \mathbf{x}_{15}^*$ be this bootstrap sample. A typical bootstrap sample might be $\mathbf{x}_5, \mathbf{x}_4, \mathbf{x}_5, \mathbf{x}_1, \mathbf{x}_1, \mathbf{x}_2, \mathbf{x}_{11}, \mathbf{x}_8, \mathbf{x}_1, \mathbf{x}_9, \mathbf{x}_7, \mathbf{x}_5, \mathbf{x}_{12}, \mathbf{x}_{11}$. Here $\mathbf{x}_3, \mathbf{x}_6, \mathbf{x}_{10}, \mathbf{x}_{13}, \mathbf{x}_{14}$ and $\mathbf{x}_{15}$ do not appear at all. The vectors $\mathbf{x}_2, \mathbf{x}_4, \mathbf{x}_7, \mathbf{x}_8, \mathbf{x}_9$ and $\mathbf{x}_{12}$ appear once each; $\mathbf{x}_{11}$ appears twice; and $\mathbf{x}_1$ and $\mathbf{x}_5$ both appear three times. From this bootstrap sample an estimate of R_1, say R_1^*, can be obtained. This resampling and estimation step is then repeated a large number of times (say, B times, where B might be as large as 1000 or more) to produce B separate estimates of R_1^*. These B values provide information on the bootstrap distribution. The rationale behind the method is that the bootstrap distribution of R_1^* will mimic the sampling distribution of R_1. The B values of R_1^* can be used to provide an estimate of the standard error of R_1

Table 5.3 Distribution of 1000 bootstrap estimates of R (from Table 1.2)

Value of R	Frequency	Cumulative frequency
<0.80	1	1
≥0.80 and <0.81	1	2
≥0.81 and <0.82	0	2
≥0.82 and <0.83	1	3
≥0.83 and <0.84	0	3
≥0.84 and <0.85	0	3
≥0.85 and <0.86	3	6
≥0.86 and <0.87	0	6
≥0.87 and <0.88	0	6
≥0.88 and <0.89	3	9
≥0.89 and <0.90	10	19
≥0.90 and <0.91	4	23
≥0.91 and <0.92	17	40
≥0.92 and <0.93	25	65
≥0.93 and <0.94	62	127
≥0.94 and <0.95	98	225
≥0.95 and <0.96	124	349
≥0.96 and <0.97	165	514
≥0.97 and <0.98	182	696
≥0.98 and <0.99	181	877
≥0.99 and <1.00	99	976
1.00 (Heywood cases)	24	1000

(using the standard deviation of the values of R_1^*) or say, a 95% confidence interval from the bootstrap cumulative distribution function.

Table 5.3 summarizes the results of taking 1000 bootstrap samples of size 15 from the 15 triplets given in Table 1.2. For each of the 1000 bootstrap samples the elements of the sample covariance matrix were calculated, the parameters β_1^* and σ_1^{*2} estimated (fixing $\sigma^2 = 1$) and finally R_1^* calculated using Equation (5.12). The 1000 values of R_1^* were then grouped into intervals as indicated in Table 5.3. A histogram illustrating the results in Table 5.3 is shown in Figure 5.1. An approximate 95% confidence interval for R_1 is 0.90–0.99.

The generation of a bootstrap distribution for a reliability coefficient can be computationally very expensive depending on the method of estimation used for each of the bootstrap samples. Very often, however, it is easy to carry out without the use of sophisticated software or the use of mainframe or mini-computers. It would be relatively straightforward, for example, to produce a simple BASIC program for analysis of the present example on a desktop microcomputer. Similar programs could be written for bootstrap distributions for such statistics as Cronbach's α coefficient or the various forms of Cohen's κ.

If all that is needed is an estimate of the standard error of a reliability coefficient, or other sample statistic, then a *jackknife* estimate would be sufficient. The jackknife estimate of standard error was introduced by Tukey in 1958 (see Mosteller and Tukey, 1977). The method is related to the bootstrap, but is computationally a lot quicker. As above, let R_1 be the reliability coefficient computed from the complete sample of 15 string elements in Table 1.2. Let $R_{1(-k)}$ be the value R_1 when the kth triplet of observations is

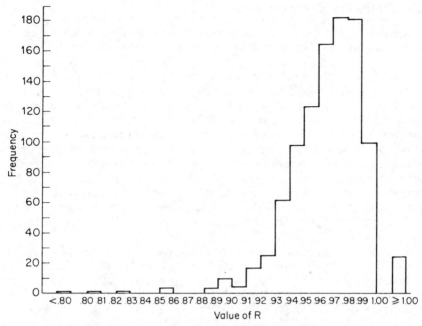

Fig. 5.1 Distribution of 1000 bootstrap estimates of *R* (see Table 5.3).

deleted from the sample. The kth *pseudovalue* is calculated as

$$R_1^{(k)} = nR_1 - (n - 1)R_{1(-k)} \tag{5.13}$$

where here $n = 15$. The jackknife estimate of R_1 is

$$R_1^{(\cdot)} = \frac{1}{n} \sum_{k=1}^{n} R_1^{(k)} \tag{5.14}$$

and its variance is given by

$$s^2(R_1^{(\cdot)}) = \frac{1}{n(n-1)} \sum (R_1^{(k)} - R_1^{(\cdot)})^2 \tag{5.15}$$

The standard error of R_1 is then simply estimated from the square root of this variance. It will be left as an exercise for the reader to calculate this standard error for Table 1.2. Further use of jackknife standard errors will be illustrated in Chapters 6 and 7.

Before leaving the data in Table 1.2 one further point should be noted. The bootstrap distribution in Table 5.3 includes values of $R_1^* = 1$. These correspond to bootstrap samples where the variance of the measurement errors has been estimated as 0 or even given a negative estimate. In the factor analysis literature these situations are referred to as *Heywood cases*. Here they can simply be regarded as the result of sampling fluctuations but the reader is warned not to assume that this will always be a sensible explanation. Negative variance estimates will be discussed in greater detail in the following section.

5.3   Pairwise comparisons

The data in Table 1.1 contain repeated measurements of the weights of several packets of potatoes. The analysis of these results, analogous to those obtained in a simple test–retest reliability study on psychology, is particularly straight-forward since it is reasonable to assume that the variance of the measurement errors does not change from the time of the first weight to the time of the second (see Equation (1.12)). Table 5.4 shows the comparison of weights obtained by the use of two *different* kitchen scales. Here it is not necessarily appropriate to assume that the precisions of the two instruments are the same. As well as the raw data, Table 5.4 gives several summary statistics for these data. Measures of association between or the consistency of the pairs of measurements are given by the product-moment correlation, intra-class correlation and Cronbach's α coefficient. For all practical purposes they can be regarded as being identical. Their sampling distributions and standard errors can be investigated through the use of bootstrap or jackknife techniques (see Exercise 5.2).

The above coefficients, however, give only some idea of the joint perform-ance of the two instruments. Obviously, we would like to be able to obtain information on relative biases and also on the precisions of the two instruments separately. We will approach the problem by considering the method described in Section 5.1. Applying Equation (5.1) to Scale A and to Scale B, we obtain

$$X_{iA} = \alpha_A + \beta_A \mu_i + e_{iA}$$

and   (5.16)

$$X_{iB} = \alpha_B + \beta_B \mu_i + e_{iB}$$

Table 5.4 Comparison of two kitchen scales: Data set 1 (weights in grammes)

Item	Scale A	Scale B	A − B	A + B
1	135	165	− 30	300
2	940	910	30	1 850
3	1 075	1 060	15	2 135
4	925	925	0	1 850
5	2 330	2 290	40	4 620
6	2 870	2 850	20	5 720
7	1 490	1 425	65	2 915
8	2 110	2 050	60	4 160
9	650	630	20	1 280
10	1 380	1 370	10	2 750
11	970	1 000	− 30	1 970
12	1 000	1 000	0	2 000
13	1 640	1 575	65	3 215
14	345	345	0	690
15	310	320	− 10	630
Mean	1 211.33	1 194.33	17.00	2 405.67
Variance	600 976.67	572 403.10	946.00	2 345 813.81

Corr(A, B):	0.9995
Cov(A, B):	586 222.65
Intra-class correlation:	0.9989
Alpha coefficient:	0.9996

The expected variances of the two scales are given by

$$\text{Var}(X_{iA}) = \beta_A^2 \sigma^2 + \sigma_A^2$$

and   (5.17)

$$\text{Var}(X_{iB} = \beta_B^2 \sigma^2 + \sigma_B^2$$

Their expected covariance is given by

$$\text{Cov}(X_{iA}, X_{iB}) = \beta_A \beta_B \sigma^2 \tag{5.18}$$

The definition of the random variables and parameters in (5.16)–(5.18) is exactly analogous to those in (5.1). If we proceed to attempt to estimate the model parameters by equating the expected dispersion matrix with that observed, we then obtain

$$S_{AA} = \beta_A^2 \sigma^2 + \sigma_A^2$$

$$S_{BB} = \beta_B^2 \sigma^2 + \sigma_B^2$$

and   (5.19)

$$S_{AB} = \beta_A \beta_B \sigma^2$$

Here we have three equations describing the mutual interrelationships of five parameters. The model is not identified. Furthermore, even if we introduce the constraint $\beta_A = 1$ or $\sigma^2 = 1$, the model is still not identified. We need a further constraint. It is necessary to assume that $\beta_A = \beta_B$ ($= 1$, say). We then obtain

$$S_{AA} = \sigma^2 + \sigma_A^2$$

$$S_{BB} = \sigma^2 + \sigma_B^2$$

and   (5.20)

$$S_{AB} = \sigma^2$$

The unique estimates for σ^2, σ_A^2 and σ_B^2 are then obtained as

$$\hat{\sigma}^2 = S_{AB}$$

$$\hat{\sigma}_A^2 = S_{AA} - S_{AB}$$

and   (5.21)

$$\hat{\sigma}_B^2 = S_{BB} - S_{AB}$$

Application of Equation (5.21) to the data in Table 5.4 now leads to another difficulty; S_{AB} is larger than S_{BB}! The estimate of σ_B^2 is negative. This is an improper solution since variances can only have values greater than or equal to zero. The estimated variance for scale A is positive, on the other hand ($\hat{\sigma}_A^2 = 14\,754.02$) and its reliability can be estimated for the ratio $S_{AB}/S_{AA} = 0.975$.

Could this unsatisfactory outcome have occurred by chance? Is there a possibility that instrument B is providing weights that are not subject to error? Or is the assumption that $\beta_A = \beta_B$ unrealistic? The reader might justifiably complain that an illustrative data set should have been chosen that would provide satisfactory estimates. The data in Table 5.4 have been deliberately chosen, however, as a warning of the potential difficulties in the interpretation of measurements provided by only two alternative measuring instruments.

In this case, prior knowledge of the two weighing scales would suggest that scale B is actually the less reliable of the two. It would certainly be silly in this case to conclude that scale B is providing error-free measurements. It is, perhaps, tempting to dismiss the negative variance estimate as merely being the result of chance. A repeat experiment using the same two scales (see Exercise 5.3) indicates, in fact that the above findings can be replicated. A more realistic conclusion is that the assumption that $\beta_A = \beta_B$ was not justified. This conclusion would then lead one to question the validity of the estimation methods for *both* σ_A^2 and σ_B^2. As data for the investigation of the *individual* characteristics of the two weighing instruments the data in Table 5.4 are practically worthless.

Moving on to another set of data on estimates of string length (Table 5.5) one can apply Equations (5.21) to the second pair of estimates (that is, to $G2$ and $G3$). Here, denoting the error variance for $G2$ and $G3$ as σ_2^2 and σ_3^2, respectively, the required estimates are

$$\hat{\sigma}_2^2 = 0.066$$

$$\hat{\sigma}_3^2 = 0.060$$

and

$$\hat{\sigma}^2 = 2.571$$

The estimate $G1$ in Table 5.5 corresponds to the first subject's (i.e. Graham's) estimates on Table 1.2. $G2$ and $G3$ are repeat estimates by the same subject made *after* seeing the results in Table 1.2. Here, then, there is a possibility that Graham has learnt to make less biased estimates following the analysis

Table 5.5 Repeat estimates of string lengths (to nearest 1/10th inch)

String	G 1	G 2	G 3	Total
A	5.0	5.5	5.1	15.6
B	3.2	3.5	4.5	11.2
C	3.6	4.4	4.8	12.8
D	4.5	4.5	4.5	13.5
E	4.0	5.1	5.1	14.2
F	2.5	2.6	3.2	8.3
G	1.7	1.8	1.9	5.4
H	4.8	4.7	4.6	14.1
I	2.4	2.3	2.3	7.0
J	5.2	5.6	6.3	17.1
K	1.2	1.1	1.2	3.5
L	1.8	1.8	1.9	5.5
M	3.4	3.6	3.7	10.7
N	6.0	6.2	6.2	18.4
O	2.2	2.1	2.3	6.6
Mean	3.433	3.653	3.840	10.927
Variance	2.121	2.637	2.631	21.645

Cov($G1, G2$): 2.315   Corr($G1, G2$): 0.979
Cov($G1, G3$): 2.239   Corr($G1, G3$): 0.948
Cov($G2, G3$): 2.571   Corr($G2, G3$): 0.976

Alpha coefficient: 0.988

in Table 1.2. There is no reason to believe, however, that Graham's three estimates are not independent conditional on the true length of the pieces of string.

If we now assume the following measurement model for all three estimates:

$$X_{i1} = \beta_1 \mu_i + e_{i1}$$

$$X_{i2} = \beta_2 \mu_i + e_{i2}$$

and

$$X_{i3} = \beta_3 \mu_i + e_{i3}$$

(5.22)

where X_{i1}, X_{i2} and X_{i3} are the estimates of the length of string i corresponding to $G1$, $G2$ and $G3$ respectively, then we can estimate the parameters of this model using the methods of the previous section. Introducing the constraint $\beta_1 = 1$, we obtain

$$\hat{\sigma}_1^2 = 0.105, \qquad \hat{\sigma}_2^2 = -0.020 \qquad \hat{\sigma}_3^2 = 0.1426$$

$$\hat{\beta}_2 = 1.148 \qquad \hat{\beta}_3 = 1.111 \qquad \hat{\sigma}^2 = 2.016$$

Once again, there is a negative variance estimate! We will return to this problem in the next section.

Typically if $G1$, $G2$ and $G3$ represented replicate measurements on 15 objects one might be prepared to assume that $\beta_1 = \beta_2 = \beta_3 = 1$. In this case we have the model

$$X_{i1} = \mu_i + e_{i1}$$

$$X_{i2} = \mu_i + e_{i2}$$

(5.23)

$$X_{i3} = \mu_i + e_{i3}$$

In this situation the parameter $\sigma^2 = \text{Var}(\mu_i)$ is over-identified. We can obtain three unique estimates; that is, S_{12}, S_{13} and S_{23}. If we could obtain a 'best' single estimator of σ^2, then we could get single unique estimates for the remaining variances; that is σ_1^2, σ_2^2 and σ_3^2. This problem, however, will be left for discussion in the following section.

Returning to Equations (5.23) we can obtain the following differences:

$$Y_{i1} = X_{i1} - X_{i2} = e_{i1} - e_{i2}$$

$$Y_{i2} = X_{i1} - X_{i3} = e_{i1} - e_{i3}$$

and

(5.24)

$$Y_{i3} = X_{i2} - X_{i3} = e_{i2} - e_{i3}$$

These are tabulated in Table 5.6. It follows from the usual model assumptions and from Equations (5.24) that

$$\text{Var}(Y_{i1}) = \sigma_1^2 + \sigma_2^2$$

$$\text{Var}(Y_{i2}) = \sigma_1^2 + \sigma_3^2$$

and

(5.25)

$$\text{Var}(Y_{i3}) = \sigma_2^2 + \sigma_3^2$$

Table 5.6 Paired differences for string length estimates $(G1, G2, G3)$ from Table 5.5

String	$G1 - G2$	$G1 - G3$	$G2 - G3$
A	−0.5	−0.1	0.4
B	−0.3	−1.3	−1.0
C	−0.8	−1.2	−0.4
D	0.0	0.0	0.0
E	−1.1	−1.1	0.0
F	−0.1	−0.7	−0.6
G	−0.1	−0.2	−0.1
H	0.1	0.2	0.1
I	0.1	0.1	0.0
J	−0.4	−1.1	−0.7
K	0.1	0.0	−0.1
L	0.0	−0.1	−0.1
M	−0.2	−0.3	−0.1
N	−0.2	−0.2	0.0
O	0.1	−0.1	−0.2
Mean	−0.22	−0.41	−0.19
Variance	0.126	0.272	0.124

$\hat{\sigma}_1^2 = 0.5(0.126 + 0.272 - 0.124) = 0.137$

$\hat{\sigma}_2^2 = 0.5(0.126 - 0.272 + 0.124) = -0.011$

$\hat{\sigma}_3^2 = 0.5(-0.126 + 0.272 + 0.124) = 0.135$

Therefore

$$\hat{\sigma}_1^2 = \tfrac{1}{2}[\text{Var}(Y_{i1}) + \text{Var}(Y_{i2}) - \text{Var}(Y_{i3})]$$

$$\hat{\sigma}_2^2 = \tfrac{1}{2}[\text{Var}(Y_{i1}) + \text{Var}(Y_{i3}) - \text{Var}(Y_{i2})]$$

and

$$\hat{\sigma}_3^2 = \tfrac{1}{2}[\text{Var}(Y_{i2}) + \text{Var}(Y_{i3}) - \text{Var}(Y_{i1})]$$

(5.26)

These expressions can be used to provide the required estimates of precision. They are given in Table 5.6. Note that the estimate of σ_2^2 is still improper. Perhaps a more suitable model would be one in which the βs (relative biases) are free to vary for one estimate to another but that the precisions are assumed to be equal. Here the error variance would be over-identified (see Section 5.4). One could also try constraining the βs to be equal as well as assuming equality of error variances. In each case one should check the identifiability of each parameter using the methods described in the present section and in Section 5.1. But one also needs a method of model fitting and parameter estimation which is more general than what has been introduced so far. This is the subject of the following section.

5.4 Maximum likelihood estimation

In specifying a measurement model in the two sections above we have derived a matrix of expected values for the variances and covariances of the initial measurements. This will be denoted by Σ. The corresponding sample dispersion matrix will be denoted by S. So far, to estimate the parameters of the measurement model we have simply equated the expected and observed sample dispersion matrices and solved the resulting set of simultaneous equations. It has been pointed out that the model must always be identified in order for unique parameter estimates to be obtainable but, if the model is *over-identified*, there will be *at least two unique solutions for at least one of the unknown parameters*. In the latter situation we need a criterion to allow us to select the 'best' estimates for the over-identified parameters. There are several optimality criteria that could be used. One of the simplest is the *unweighted least squares* criterion specified by minimizing

$$F_{ULS} = \sum_{i=1}^{p} \sum_{h=1}^{p} (S_{ih} - \sigma_{ih})^2$$

$$= \text{tr}(\mathbf{S} - \mathbf{\Sigma})^2 \qquad (5.27)$$

Here, p is the number of measurements taken on each subject and S_{ih} and σ_{ih} are the observed and expected covariances between measurements i and h, respectively. The symbol 'tr' is the trace operator indicating the sum of the diagonal elements of a matrix. A second criterion that might be used is that involving the minimisation of

$$F_{GLS} = \text{tr}\{(\mathbf{S} - \mathbf{\Sigma})\mathbf{S}^{-1}\}^2 \qquad (5.28)$$

F_{GLS} is the *generalised least squares* criterion (see Jöreskog and Goldberger, 1972). A third criterion involves minimizing

$$F_{ML} = \text{tr}(\mathbf{S}\mathbf{\Sigma}^{-1}) + \log|\mathbf{\Sigma}| - \log|\mathbf{S}| - q \qquad (5.29)$$

where q is the number of latent variables or factors. If the measurements can be assumed to be distributed following the *multivariate normal distribution*, then it can be shown that the estimates obtained by minimizing F_{ML} are the *maximum likelihood* estimates (see Lawley and Maxwell, 1971). In the examples discussed in this chapter it will always be assumed that the observations are multivariate normal and therefore F_{ML} will be used as the optimality criterion throughout. The use of a maximum likelihood factor analysis program such as LISREL (see Appendix 4) has the advantage of providing a large-sample goodness-of-fit statistic based on the log-likelihood ratio criterion and also the standard errors of the parameter estimates. Details of the theoretical properties of maximum likelihood estimation are beyond the scope of this text, however, and the interested reader is referred to Lawley and Maxwell (1971) or Bartholomew (1987) for further information. One can usually also obtain covariances between each pair of parameter estimates as part of the output from a maximum-likelihood fitting procedure and can therefore use the delta technique (see Appendix 3) to estimate standard errors of statistics derived from these parameter estimates (reliability estimates, for example).

Returning to the measurement model for the estimates of string lengths provided by Graham, Brian and David (Equation (5.1)), it will be seen that, so far, we have not allowed for the estimation of the intercept terms (the αs). This problem can be solved by considering the expected moments about zero rather than the measurement means. From Equation (5.1) we have

$$E(X_{ik}^2) = E[\alpha_k + \beta_k \mu_i + e_{ik}]^2 \qquad (5.30)$$

and

$$E(X_{ik} X_{im}) = E\{(\alpha_k + \beta_k \mu_i + e_{ik})(\alpha_m + \beta_m \mu_i + e_{im})\} \quad (5.31)$$

These define the expected *moments matrix*. If we now augment the three measurements by a fourth one, X_4, which is set equal to one for all pieces of string, then

$$E(X_4^2) = 1 \quad (5.32)$$

and

$$E(X_{ik} X_4) = E(X_{ik})$$

$$= \alpha_k + \beta_k \bar{\mu} \quad (5.33)$$

for $k = 1$, 2 or 3. Equations (5.30)–(5.33) define an *augmented moments matrix*. If we now introduce constraints to enable the model to be identified, then one of the alternatives is to introduce the constraints $\bar{\mu} = 0$ and $\sigma_\mu^2 = 1$. The expected augmented moments matrix obtained after introducing these two constraints is shown in Table 5.7(a). If, instead of the constraints $\bar{\mu} = 0$ and $\sigma_\mu^2 = 1$, we let $\alpha_1 = 0$ and $\beta_1 = 1$, then the expected moments matrix is given by Table 5.7(b). The parameter estimates can be obtained in both cases by equating observed and expected moments and then solving the resulting simultaneous equations. Alternatively a maximum likelihood program can be used using the criterion given in (5.29), but in this case replacing the observed and expected dispersion matrices by observed and expected augmented moments matrices, respectively. Of course, unweighted or generalized least squares procedures could also be used in a similar way. The observed moments for the data on string lengths are given in Table 5.8. Maximum likelihood estimates for the parameter when $\bar{\mu} = 0$ and $\sigma_\mu^2 = 1$ are given in Table 5.9(a). Some of them differ from the solutions given by Equations (5.6) and (5.9) simply because of rounding errors introduced by calculation of the elements of the covariance matrix in Table 5.1 using a pocket calculator. They should, of

Table 5.7 Expected augmented moments matrices for string lengths in Table 1.2

(a) When $\bar{\mu} = 0$ and $\sigma_\mu^2 = 1$

	Graham	Brian	David	X4
Graham	$\alpha_1^2 + \beta_1^2 + \sigma_1^2$			
Brian	$\alpha_2 \alpha_1 + \beta_2 \beta_1$	$\alpha_2^2 + \beta_2^2 + \sigma_2^2$		
David	$\alpha_3 \alpha_1 + \beta_3 \beta_1$	$\alpha_3 \alpha_2 + \beta_3 \beta_2$	$\alpha_3^2 + \beta_3^2 + \sigma_3^2$	
X4	α_1	α_2	α_3	1

(b) When $\alpha_1 = 0$ and $\beta_1 = 1$

	Graham	Brian	David	X4
Graham	$\bar{\mu}^2 + \sigma_\mu^2 + \sigma_1^2$			
Brian	$\bar{\mu}(\alpha_2 + \beta_2 \bar{\mu}) + \beta_2^2 \sigma_\mu^2$	$(\alpha_2 + \beta_2 \bar{\mu})^2 + \beta_2^2 \sigma_\mu^2 + \sigma_2^2$		
David	$\bar{\mu}(\alpha_3 + \beta_3 \bar{\mu}) + \beta_3^2 \sigma_\mu^2$	$(\alpha_3 + \beta_3 \bar{\mu})(\alpha_2 + \beta_2 \bar{\mu}) + \beta_3 \beta_2 \sigma_\mu^2$	$(\alpha_3 + \beta_3 \bar{\mu})^2 + \beta_3^2 \sigma_\mu^2 + \sigma_3^2$	
X4	$\bar{\mu}$	$\alpha_2 + \beta_2 \bar{\mu}$	$\alpha_3 + \beta_3 \bar{\mu}$	1

Table 5.8 Observed augmented moments matrix for string data in Table 1.2

	Graham	Brian	David	X4
Graham	13.765			
Brian	13.881	14.222		
David	15.330	15.558	17.210	
X4	3.433	3.427	3.767	1.000

course, be identical. Table 5.9(a) also provides standard errors of the estimates but, in this case, because the model is just identified, there are no degrees of freedom for a goodness-of-fit test. The model fits perfectly.

The estimates of the αs in Table 5.9(a) are clearly interpretable as means. In the model involving the alternative constraints ($\alpha_1 = 0$ and $\beta_1 = 1$) the αs are interpreted as intercepts. That is, α_k is the expected value of X_{ik} when $\mu_i = 0$. In the case of the three different sets of string 'measurements' a further model of interest is one in which all the αs are constrained to be zero. The maximum likelihood estimates for this case (but using the constraint $\sigma_\mu^2 = 1$) are shown in Table 5.9(b). Note that here there are two degrees of freedom available to test the fit of the model (these having arisen by setting $\alpha_2 = 0$ and $\alpha_3 = 0$). Comparison of chi-square, the log-likelihood ratio statistic, with a $\chi_{[2]}^2$ variate suggests that the fit of the reduced model is perfectly adequate.

Table 5.9 Maximum likelihood estimates for string lengths data in Table 1.2 (analysis of augmented moments matrix)

(a) Full model; $\bar{\mu} = 0$, $\sigma_\mu^2 = 1$

$\hat{\beta}_1 = 1.384$ (s.e. 0.261)	$\hat{\alpha}_1 = 3.433$ (s.e. 0.363)
$\hat{\beta}_2 = 1.529$ (s.e. 0.296)	$\hat{\alpha}_2 = 3.427$ (s.e. 0.406)
$\hat{\beta}_3 = 1.733$ (s.e. 0.318)	$\hat{\alpha}_3 = 3.767$ (s.e. 0.449)
$\hat{\sigma}_1^2 = 0.064$ (s.e. 0.034)	
$\hat{\sigma}_2^2 = 0.141$ (s.e. 0.060)	
$\hat{\sigma}_3^2 = 0.018$ (s.e. 0.040)	

Chi-square = 0 with 0 d.f.

(b) Constrained model; ($\sigma_\mu^2 = 1$; $\alpha_1 = \alpha_2 = \alpha_3 = 0$)

$\bar{\mu} = 2.263$ (s.e. 0.491)	$\hat{\sigma}_1^2 = 0.075$ (s.e. 0.042)
$\hat{\beta}_1 = 1.495$ (s.e. 0.277)	$\hat{\sigma}_2^2 = 0.130$ (s.e. 0.058)
$\hat{\beta}_2 = 1.517$ (s.e. 0.282)	$\hat{\sigma}_3^2 = 0.043$ (s.e. 0.043)
$\hat{\beta}_3 = 1.674$ (s.e. 0.309)	

Chi-square = 4.50 with 2 d.f.

5.5 Confirmatory factor analysis

Essentially, confirmatory factor analysis (CFA) involves the specification of alternative measurement models for a given set of data, estimation of the required parameter using maximum likelihood or one of the other fitting criteria, and testing the adequacy of fit of the model. Quite often there will be a single measurement model that is the obvious choice for a set of measurements but usually one will be interested in comparison of the fit of a series of *nested* models. For example, one might fit a single-factor model to a set of congeneric tests and then proceed to test for equality of the factor loadings. This is equivalent to asking whether the tests are τ-equivalent. The τ-equivalence model is nested within the congeneric tests model; it is arrived at by imposing constraints on the parameters of the more general congeneric tests model. Similarly the parallel tests model is a special case of both the τ-equivalence and congeneric tests model. It is obtained by constraining the error variances of the τ-equivalence model to be equal. As is usual with a series of nested alternatives, one can test the fit of each of the models separately using the likelihood ratio criterion and one can also look at improvement or deterioration in the fit as one passes from one model to another within the nested series of alternatives. The aim of fitting alternative models is, of course, to find the *simplest* model which appears to be consistent with the data.

Quite often the parameter estimates for the simpler models will be 'better behaved'. Their standard errors are often smaller and it is not uncommon to find that improper estimates such as negative variances disappear after the introduction of a few sensible constraints. Returning to the string lengths reported in Table 5.5, for example, we found that fitting the single factor model (Equation (5.22)) led to a negative error variance for $G2$. The assumption of τ-equivalence (Equation (5.23)) did not, however, lead to a better outcome. An alternative procedure is to assume that the systematic biases for the three sets of string lengths ($G1$, $G2$ and $G3$) may differ as a result of learning but that the random error variances remain more or less the same. These assumptions imply a measurement model of the same form as Equation (5.22) but with the constraints $\sigma_1^2 = \sigma_2^2 = \sigma_3^2 \ (=\sigma_e^2$, say). We might now hope that this sample would fit the data and that the estimate of the common error variance (σ_e^2) would be positive. It will be left as an exercise for the reader to check that this model is identified. Maximum likelihood estimates of the free parameters are the following:

$$\hat{\beta}_2 = 1.128, \qquad\qquad \hat{\beta}_3 = 1.117,$$

$$\hat{\sigma}_e^2 = 0.077 \quad \text{and} \quad \hat{\sigma}^2 = 2.034$$

The likelihood-ratio statistic is 6.34 with 2 d.f. The fit is not terribly good, although there are now no improper parameter estimates. If we now constrain β_2 and β_3 to be both equal to β_1 (i.e. $\beta_1 = \beta = 1$), then the resulting chi-square is 9.80 with 4 d.f. The difference between the two models can be assessed from $9.80 - 6.34$ with 2 d.f. There is clearly very little evidence of learning here, but this might be simply due to lack of power of statistical tests based on such a small sample of observations. If we accept that the three measures of string length are indeed parallel tests, then there are only two free parameters to estimate. The maximum likelihood estimates are $\hat{\sigma}^2 = 2.376$ and $\sigma_e^2 = 0.087$.

Now consider the measurements given in Table 5.10. These are derived from computer-aided tomographic scans (CAT scans) of the heads of 50 psychiatric patients (see Turner *et al.*, 1986). The primary aim of these scans was to determine the size of the brain ventricle relative to that of the patient's skull (the ventricle–brain ratio or VBR = (ventricle size/brain size) × 100). For given scan or 'slice' the VBR was determined from measurements of the perimeter of the patient's ventricle together with the perimeter of the inner surface of the skull. These measurements were made using either (a) a hand-held planimeter on a projection of the X-ray image, or (b) from an automated pixel count based on the image displayed on a television screen. Table 5.10 gives the logged VBRs for single scans from 50 patients. The first two columns correspond to repeated determinations based on pixel counts and the second two columns correspond to repeated determinations based on the use of a planimeter. Summary statistics for these data are given in Table 5.11. It will be left to the interested reader to produce the six possible two-way scatter plots for these data.

Table 5.10 CAT scan data (logged VBRs) on 50 patients

PIX1	PIX3	PLAN1	PLAN3	PIX1	PIX3	PLAN1	PLAN3
1.79	1.77	2.05	2.13	2.33	2.37	2.24	2.03
0.00	0.00	1.72	1.28	1.22	1.19	1.63	1.76
1.53	1.55	1.93	1.79	1.63	1.39	1.55	1.53
1.57	1.57	2.16	1.96	1.87	1.84	2.12	2.30
1.65	1.70	2.27	1.95	1.19	1.10	1.63	1.34
2.05	2.12	2.53	2.17	0.34	0.34	1.46	0.96
1.59	1.65	1.79	1.67	1.19	1.25	1.87	1.41
1.03	1.03	1.87	1.48	1.53	1.53	1.79	1.84
0.69	0.74	1.57	1.57	1.63	1.65	2.33	1.84
1.69	1.79	1.39	1.39	0.83	0.88	1.39	1.16
1.50	1.55	1.89	1.84	1.10	1.10	1.96	1.53
1.74	1.72	2.39	2.26	1.76	1.76	2.40	2.30
1.50	1.63	1.67	1.72	1.41	1.44	2.09	1.89
0.74	0.74	1.57	1.39	0.92	0.96	1.39	1.41
1.67	1.69	2.30	2.25	1.63	1.65	2.22	1.89
1.61	1.59	2.03	1.93	0.74	0.79	1.67	1.34
1.03	0.99	1.19	1.70	0.74	0.79	2.03	1.46
0.88	0.96	1.13	0.41	1.36	1.36	2.26	2.12
1.25	1.28	1.63	1.22	1.28	1.31	1.69	1.63
1.79	1.77	1.93	2.03	2.30	2.29	2.30	2.50
1.84	1.89	1.89	1.50	1.39	1.34	2.01	1.50
1.22	1.22	1.63	2.03	1.16	1.16	1.57	1.59
1.90	1.99	1.70	1.96	0.69	0.69	2.08	1.55
2.91	2.93	2.82	2.84	1.95	1.95	2.13	2.09
1.19	1.10	0.53	0.99	1.57	1.55	1.69	1.13

Looking at Table 5.11 we can see that the VBRs based on the pixel count are very highly correlated (0.994). Those based on the planimeter, however, are much less consistent (correlation 0.785). The means of the pixel VBRs tend to be lower than the planimetry estimates but, on the other hand, their variances are higher. The correlation between a pixel VBR and a planimetry estimate is reasonably high but, perhaps a little disappointing. These data suggest that the pixel method is considerably more 'reliable' than the older planimetry-based one. The raw data in Table 5.10, however, hint that this might be an over-simplification. The second subject in Table 5.10, for example, has a value of 0 for the logged VBR based on the pixel count (corresponding to a VBR of 1). This value is consistently lower than the corresponding planimetry-based determinations.

Table 5.11 Summary statistics for the logged VBR data shown in Table 5.10

(a) Covariance matrix

	PIX1	PIX3	PLAN1	PLAN3
PIX1	0.272			
PIX3	0.272	0.276		
PLAN1	0.123	0.126	0.163	
PLAN3	0.168	0.168	0.139	0.192

(b) Correlation matrix

	PIX1	PIX3	PLAN1	PLAN3
PIX1	1			
PIX3	0.994	1		
PLAN1	0.585	0.595	1	
PLAN3	0.736	0.730	0.785	1

(c) Means

PIX1	PIX3	PLAN1	PLAN3
1.403	1.412	1.862	1.711

Looking down the 50 patients we can see other examples where both of the pixel estimates are considerably lower than the planimeter-based ones. It appears that, although the method is very consistent, the pixel count occasionally leads to a gross error. On the other hand, although they are more erratic, the planimeter-based measurements appear to be much less prone to these gross errors.

One possible measurement model for these data has the following form:

$$X_{ijk} = \tau_i + d_{ik} + e_{ijk} \tag{5.34}$$

Here X_{ijk} is the logged VBR for the jth replicate ($j = 1, 2$) of the ith subject ($i = 1$ to 50) using measurement method k ($k = 1, 2$). The e_{ijk} are independent random measurement errors as before and the d_{ik} are components of error that are common to both replicates using method k. The 'error' terms (e_{ijk} and d_{ik}) are assumed to be uncorrelated and also uncorrelated with the 'true' value, τ_i. If $k = 1$ for the pixel measure and $k = 2$ from the planimetry measure, then the reliability of the two methods can be defined by

$$R_1 = \frac{\sigma^2}{\sigma^2 + \sigma_{d1}^2 + \sigma_{e1}^2}$$

and

$$R_2 = \frac{\sigma^2}{\sigma^2 + \sigma_{d2}^2 + \sigma_{e2}^2} \tag{5.35}$$

respectively. Here $\sigma^2 = \mathrm{Var}(\tau_i)$, $\sigma_{dk}^2 = \mathrm{Var}(d_{ik})$ where $k = 1$ or 2, and $\sigma_{ek}^2 = \mathrm{Var}(e_{ijk})$ where $k = 1$ or 2. Equation (5.34) is equivalent to a factor analysis model in which there are three specified orthogonal factors plus measurement error. The first factor is common to all four measurements, the second is common to the two pixel measurements, and the third is common to the two planimetry measures. If all of the non-zero factor loadings are constrained to be equal to 1 then, if the factors are called f_1, f_2 and f_3 respectively, we have

$$\text{Var}(f_1) = \sigma^2$$
$$\text{Var}(f_2) = \sigma^2_{d_1}$$

and

$$\text{Var}(f_3) = \sigma^2_{d_2}$$

(5.36)

The expected covariance matrix for the four VBR measures is shown in Table 5.12(a). The model is clearly over-identified and the parameters can be estimated using a program such as LISREL.

Table 5.12 Expected covariance matrices for logged VBR measurements

(a) Model: $X_{ijk} = \tau_i + d_{ik} + e_{ijk}$

	PIX1	PIX3	PLAN1	PLAN3
PIX1	$\sigma^2 + \sigma^2_{d_1} + \sigma^2_{e_1}$			
PIX3	$\sigma^2 + \sigma^2_{d_1}$	$\sigma^2 + \sigma^2_{d_1} + \sigma^2_{e_1}$		
PLAN1	σ^2	σ^2	$\sigma^2 + \sigma^2_{d_2} + \sigma^2_{e_2}$	
PLAN3	σ^2	σ^2	$\sigma^2 + \sigma^2_{d_2}$	$\sigma^2 + \sigma^2_{d_2} + \sigma^2_{e_2}$

(b) Model: $X_{ij1} = \mu_i + e_{ij1}$; $X_{ij2} = \eta_i + e_{ij2}$

	PIX1	PIX3	PLAN1	PLAN3
PIX1	$\sigma^2_\mu + \sigma^2_{e_1}$			
PIX3	σ^2_μ	$\sigma^2_\mu + \sigma^2_{e_1}$		
PLAN1	$\rho\sigma_\eta\sigma_\mu$	$\rho\sigma_\eta\sigma_\mu$	$\sigma^2_\eta + \sigma^2_{e_2}$	
PLAN3	$\rho\sigma_\eta\sigma_\mu$	$\rho\sigma_\eta\sigma_\mu$	σ^2_η	$\sigma^2_\eta + \sigma^2_{e_2}$

Instead of estimating these parameters directly, however, we will approach the problem by specification of a CFA model that is superficially quite different. Consider a model for the pixel measures. Here we have

$$X_{ij1} = \mu_i + e_{ij1}$$

(5.37)

where μ_i is a 'true' pixel score with variance σ^2_μ. As before we assume that the errors are uncorrelated with the true scores and that their variance is σ^2_{ei}. Similarly, from the planimetry measures, we have

$$X_{ij2} = \eta_i + e_{ij2}$$

(5.38)

where η_i is a 'true' planimetry score with variance σ^2_η. The errors are again uncorrelated with variance $\sigma^2_{e_2}$. Obviously μ_i and η_i are correlated, and we represent this correlation with the parameter ρ. Together with ρ, Equations (5.37) and (5.38) define a two-common-factors model that can again be fitted through the use of a program such as LISREL. The expected covariance matrix for this model is shown in Table 5.12(b). The model is over-identified and the maximum likelihood parameter estimates are given in Table 5.13.

Table 5.13 Parameter estimates for the VBR models

$\hat{\sigma}_\mu^2 = 0.273$	$\hat{\sigma}_\eta^2 = 0.138$
$\sigma_{e_1}^2 = 0.002$	$\sigma_{e_2}^2 = 0.039$
$\hat{\rho} = 0.753$	
$\hat{\sigma}^2 = 0.146$	
$\hat{\sigma}_{d_1}^2 = 0.126$	$\hat{\sigma}_{d_2}^2 = -0.008$
$\hat{R}_1 = 0.531$	$\hat{R}_2 = 0.825$
Chi-square $= 8.09$ with 5 d.f.	

The parameters of Equation (5.34) can now easily be obtained by equating the elements of the two covariance matrices in Table 5.12. For example,

$$\hat{\sigma}^2 = \hat{\rho}\hat{\sigma}_\mu\hat{\sigma}_\eta$$

$$\hat{\sigma}_{d_1}^2 = \hat{\sigma}_\mu^2 - \hat{\rho}\hat{\sigma}_\mu\hat{\sigma}_\eta \qquad (5.39)$$

and

$$\hat{\sigma}_{d_2}^2 = \hat{\sigma}_\eta^2 - \hat{\rho}\hat{\sigma}_\mu\hat{\sigma}_\eta$$

From these we obtain the estimates of the reliabilities of the two methods given by

$$\hat{R}_1 = \frac{\hat{\rho}\hat{\sigma}_\mu\hat{\sigma}_\eta}{\hat{\sigma}_\mu^2 + \hat{\sigma}_{e_1}^2}$$

$$\hat{R}_2 = \frac{\rho\hat{\sigma}_\mu\hat{\sigma}_\eta}{\hat{\sigma}_\eta^2 + \hat{\sigma}_{e_2}^2} \qquad (5.40)$$

The results are shown in Table 5.13.

The analyses of the CAT scan data in Table 5.10 have emphasised again the way in which the idea of 'reliability' is context dependent. On their own the two pixel measures would indicate that the automated method of estimating VBRs is very highly reliable (test–retest correlation of 0.992). The introduction of replicate planimetry measures, however, forces one to reconsider this appraisal. The reliability defined by (5.35) is very much lower ($\hat{R}_1 = 0.531$). Examination of the raw data would suggest that the latter measure of reliability is a much better indicator of the usefulness of the pixel measure.

The interesting parameter in these models is the correlation between μ_i and η_i (that is, ρ). This can be interpreted as a measure of the 'validity' of the automated pixel measure compared with the planimetry method (or vice versa). The standard error of its estimate obtained through the use of LISREL is 0.073. The estimate of ρ is clearly significantly below 1. One can use bootstrap methods to investigate the sampling characteristics of this estimate (see Section 5.2). Returning to Table 5.12(b) it can be seen that there are two possible solutions for ρ that can be obtained by equating observed with expected covariances for the four measures. These are

$$\hat{\rho}_1^2 = \frac{\text{Cov(PIX1, PLAN1) Cov(PIX3, PLAN3)}}{\text{Cov(PIX1, PIX3) Cov(PLAN1, PLAN3)}}$$

and

$$\hat{\rho}_2^2 = \frac{\text{Cov(PIX1, PLAN3) Cov(PIX3, PLAN1)}}{\text{Cov(PIX1, PIX3) Cov(PLAN1, PLAN3)}} \qquad (5.41)$$

A common estimate can be obtained from

$$\hat{\rho} = \frac{\sqrt{\hat{\rho}_1^2 + \hat{\rho}_2^2}}{2}$$

$$= 0.745 \qquad (5.42)$$

An alternative (Spearman, 1910) is

$$\hat{\rho} = \sqrt{\hat{\rho}_1 \hat{\rho}_2}$$

$$= 0.745, \qquad \text{as before.} \qquad (5.43)$$

Table 5.14 summarises the results of taking 1000 bootstrap samples of size 50 from the 50 quadruplicate measurements given in Table 5.10. For each of the 1000 bootstrap samples the elements of the sample covariance matrix were calculated and the estimates of ρ, that is $\hat{\rho}_1^*$ and $\hat{\rho}_2^*$, determined. The correlation ρ^* was then estimated using Equation (5.42). The mean value of ρ^* was 0.741 with standard error of 0.073. (The mean of a second 1000 estimates of ρ^* was 0.741 with a standard error of 0.079.) The upper limit for the one-tailed 95% confidence interval for ρ is 0.85.

Table 5.14 Cumulative distribution of 1000 bootstrap estimates of ρ

Value of ρ	Cumulative frequency
≤ 0.80	788
≤ 0.81	832
≤ 0.82	862
≤ 0.83	903
≤ 0.84	927
≤ 0.85	951
≤ 0.86	966
≤ 0.87	979
≤ 0.88	984
≤ 0.89	989
≤ 0.90	991
≤ 0.95	999

The final point to be discussed in this section concerns the determination of the reliability of a composite measure derived from the four alternatives given in Table 5.10. Consider the arithmetic mean. This will have the same reliability as the sum of the four measures and, for simplicity, we will determine the reliability of the sum. Let

$$Y_i = X_{i11} + X_{i12} + X_{i21} + X_{i22} \qquad (5.44)$$

$$= (2\mu_i + e_{i11} + e_{i12}) + (2\eta_i + e_{i21} + e_{i22})$$

$$= 2(\mu_i + \eta_i) + e_{i11} + e_{i12} + e_{i21} + e_{i22} \qquad (5.45)$$

$$\text{Var}(Y_i) = 4\,\text{Var}(\mu_i + \eta_i) + 2\sigma_{e_1}^2 + 2\sigma_{e_2}^2 \qquad (5.46)$$

$$= 4\sigma_\mu^2 + 4\sigma_\eta^2 + 8\,\text{Cov}(\mu_i, \eta_i) + 2\sigma_{e_1}^2 + 2\sigma_{e_2}^2 \qquad (5.47)$$

Note that this is the sum of *all* of the elements of the expected covariance matrix given in Table 5.12(b). The reliability R_c of Y_i is therefore

$$R_c = \frac{4 \, \mathrm{Var}(\mu_i + \eta_i)}{\mathrm{Var}(Y_i)} \tag{5.48}$$

$$= \frac{\mathrm{Var}(Y_i) - 2\sigma_{e_1}^2 - 2\sigma_{e_2}^2}{\mathrm{Var}(Y_i)} \tag{5.49}$$

For the present example, $\hat{R}_c = 0.97$. The derivation of the optimally reliable composite score is discussed in Werts *et al.* (1978).

5.6   Pitfalls for the unwary

In Section 5.3 we saw that it is not possible to estimate the loadings and error variances for the common factor model when there are only two available measurements available on each subject. The factor analysis model is under-identified. At least three measurements are needed for each subject. The choice of three or more measurements does not, however, ensure that the reliability estimates obtained through the use of a factor analysis model are at all sensible. To illustrate this point, reconsider the data in Table 5.10. Suppose, for example, that the fourth measurement (PLAN3) has not been made. And, further suppose that a CFA model is fitted to the first three measurements. The model is

$$X_{ik} = \beta_k \mu_i + e_{ik} \tag{5.50}$$

where X_{ik} is the kth measurement on subject i. Here $k = 1$ for PIX1, $k = 2$ for PIX3 and $k = 3$ for PLAN1. The parameters are defined in a similar way to earlier models, with $\mathrm{Var}(\mu_i) = 1$. Now let $\beta_1 = \beta_2$ and $\sigma_1^2 = \sigma_2^2$ where σ_k^2 is the specific or error variance for the kth measurement. Maximum likelihood estimates for this constrained CFA model are

$$
\begin{aligned}
\hat{\beta}_1 &= \hat{\beta}_2 = 0.522 \\
\hat{\beta}_2 &= 0.239 \\
\hat{\sigma}_1^2 &= \hat{\sigma}_2^2 = 0.002 \\
\hat{\sigma}_3^2 &= 0.106
\end{aligned}
\tag{5.51}
$$

From these estimates we obtain an estimate of the reliability of the pixel measures of 0.993. The estimated reliability of PLAN1 is 0.350. The pixel measures appear to be excellent, the planimetry measures rather poor.

Now suppose that we have only obtained measurements PIX1, PLAN1 and PLAN3. If we now fit (5.50) with $k = 1$ for PIX1, $k = 2$ for PLAN1 and $k = 3$ for PLAN3 and constraints $\beta_2 = \beta_3$ and $\sigma_2^2 = \sigma_3^2$, then we obtain the following maximum likelihood estimates:

$$
\begin{aligned}
\hat{\beta}_1 &= 0.391 \\
\hat{\beta}_2 &= \hat{\beta}_3 = 0.372 \\
\hat{\sigma}_1^2 &= 0.119 \\
\hat{\sigma}_2^2 &= \hat{\sigma}_3^2 = 0.039
\end{aligned}
\tag{5.52}
$$

Now the estimated reliabilities from pixel and planimetry measures are 0.562 and 0.780, respectively. The relative performance of the two methods of measurement appears to have been reversed!

The problem arises because we cannot distinguish specific and error variance in these models. In the first of the two models, variation that is specific to the pixel measurements is confounded with the underlying latent trait (the common factor); that which is specific to the planimetry measures is confounded with random measurement error. In the second of the two models the specific variation characteristic of the pixel measurements is confounded with measurement error and that specific to the planimetry measures is incorporated into the common factor.

One clearly has to be very careful when designing a reliability study that might include replicate measurements using one or more of the alternative measurement techniques. Perhaps one should aim to have at least two indicators or replicate measurements for each of the different measurement techniques being used. The data in Table 5.10 provide the simplest example of this type of design. More complicated examples are found in so-called *multi-trait–multi-method studies* (Campbell and Fiske, 1959). These studies, however, are beyond the scope of this text and the interested reader is referred to Dwyer (1983) and to Sullivan and Feldman (1979) for further details.

Jaech (1979) has discussed some of the problems with the use of replicated measurements within the context of an interlaboratory precision study. Here reproducibility is analogous to specific method variance and repeatability is analogous to the random error variance (see Section 4.1). If, for example, several laboratories are each sent several standard preparations of ammonia solution and asked to make replicate measurements of the concentration of ammonia in preparation, then the following measurement model might apply:

$$X_{ijk} = \alpha_k + \beta_k \mu_i + d_{ik} + e_{ijk}$$

where X_{ijk} is the jth replicate measurement of sample i by laboratory k and α_k and β_k are parameters indicating the relative biases of laboratory k. The 'true' concentration of ammonia in sample i is represented by μ_i. The terms d_{ik} and e_{ijk} represent random measurement errors, d_{ik} being common to all measurements made on preparation i within laboratory k and e_{ijk} being random within-laboratory variation. This model is analogous to the simpler one for VBR measures (Equation (5.34)).

Jaech (1979) discusses two reasons why replicate data might provide misleading information concerning within-laboratory variability. The first is the temptation to 'clean' the data prior to the analysis. Obviously one wants to identify potential outliers in the data, but there is also the suggestion that by removing the discordant measurements the remaining replicates can be made to look more consistent than they really are. The second problem is caused by the fact that a laboratory might hold measurement conditions fixed while obtaining the replicate measurements. There are many sources of variability of measurements made within a laboratory and under the assumption that the laboratory's quality will be judged largely on the basis of the closeness of the replicate measurements the analyst may hold constant as many of the identified sources of variation as possible. This would lead to an underestimation of the within laboratory variation normally associated with repeated measurements.

Although Jaech's examples come from analytical chemistry laboratories his warnings are equally applicable for the measurement of instrument precision in other contexts. Behavioural scientists evaluating the performance of psychiatric rating scales, for example, will choose a combination of skilled interviewers and raters working under ideal conditions. Here too the measurement techniques will appear to be better than they would be during routine use.

5.7   Simultaneous factor analysis in several populations

This section is concerned with the investigation of similarities and differences in factor structures between different groups. A complementary aim is the estimation of parameters of a factor analysis model which is assumed to be common to two or more groups of measurements. The starting point of these investigations will be a covariance or moments matrix for each of, say, M groups. If a CFA model is fitted to the gth group ($g = 1, 2, ..., M$) using the maximum likelihood criterion given in Equation (5.29), then this is equivalent to the function

$$F_g = \text{tr}(S_g \Sigma_g^{-1}) + \log|\Sigma_g| - \log|S_g| - q_g \qquad (5.53)$$

where S_g and Σ_g are the observed and expected dispersion matrices, respectively. The q_g is the specified number of latent variables or factors for group g. If one assumes that the M samples are independent, the log-likelihood of all the samples is

$$\log L = \sum_{g=1}^{M} \log L_g \qquad (5.54)$$

where L_g is the likelihood for group g. Maximizing the likelihood in (5.45) is equivalent to minimizing

$$F_{\text{ML}} = \sum_{g=1}^{M} \frac{N_g}{N} F_g \qquad (5.55)$$

where N_g is the sample size of the gth group and N is the total sample size ($N = \sum_{g=1}^{M} N_g$). If one is fitting data to moments matrices rather than dispersion matrices one simply replaces S_g and Σ_g in Equation (5.53) by the respective moment matrices. Details are to be found in Jöreskog (1971b) and Jöreskog and Sörbom (1984). At the minimum F_{ML}, the log-likelihood ratio statistic will provide a test of the overall fit of the models which, in large samples will be distributed as a chi-square distribution with degrees of freedom (d) given by

$$d = \sum_{g=1}^{M} \tfrac{1}{2} q_g (q_g + 1) - t \qquad (5.56)$$

where t is the total number of independent parameters estimated in the models.

A preliminary pair of tests that might be of interest are (a) a test of the equality of dispersion matrices across groups and (b) a test of the equality of correlation matrices across the groups. Test (a) is achieved by creating a set of factors that are identical to the observed measurements, that is, there are as many common factors as there are measurements for each subject, the non-zero loadings (βs) are all unity, and the error variances are all zero. The dispersion matrix for the common factors is here identical to the dispersion matrix for the original measurements. In matrix terms, the general measurement model is

$$\underset{(n_g \times k)}{\mathbf{X}_g} = \underset{(n_g \times k)}{\boldsymbol{\mu}_g} \underset{(k \times k)}{\boldsymbol{\beta}_g} + \underset{(n_g \times k)}{\mathbf{e}_g} \tag{5.57}$$

where $\mathbf{X}_g$ is the matrix of observations for group g, $\boldsymbol{\beta}_g$ is a matrix of loadings on the k factors. The matrix of k factor scores for each of the n_g subjects is given by $\boldsymbol{\mu}_g$ and the corresponding measurement errors by $\mathbf{e}_g$. In this particular case $\boldsymbol{\beta}_g$ is a diagonal matrix and $\mathbf{e}_g$ is the null matrix (usually specified by setting all of the elements of the dispersion matrix for the errors equal to zero), so that

$$\mathbf{X}_g = \boldsymbol{\mu}_g \tag{5.58}$$

To test whether the dispersion matrix of the measurements is common to all M groups we simply constrain the dispersion matrix for the common factors to be equal across the groups.

Test (b) – equality of correlations – is achieved by use of the same model as that in Equation (5.57), but in this case the diagonal elements of $\boldsymbol{\beta}_g$ are not constrained to be unity. In addition the variances of the k common factors are constrained to be unity so that the dispersion matrix for the common factors is equivalent to the required correlation matrix. The off-diagonal elements of the correlation matrix are, of course, free and estimated from the data. As before, the matrix of error terms, $\mathbf{e}_g$, is the null matrix. The test of equality of correlation matrices across groups is obtained by again constraining the dispersion matrix for the common factors to be the same for each of the M groups.

Computer programs such as LISREL usually assume that the number of measurements and the number of common factors are the same for each of the M groups. If this is not the case one can get round this problem by the introduction of *pseudovariables* and *pseudofactors* (see Jöreskog, 1971b). A pseudovariable has unit observed variance, zero observed covariances with every other variable, zero factor loadings on each factor including any pseudofactors, and also unit unique or measurement error variance. Similarly, each pseudofactor has unit variance and zero covariance with every other factor and pseudofactor. Such pseudofactors and pseudovariables have no effect on the likelihood function whatsoever. They are simply introduced as a trick to allow the use of commonly available software in this situation.

The use of pseudovariables will be illustrated by return to CAT scan data similar to that described in Section 5.5 (see Table 5.11). Table 5.15 provides data from two samples; one of 50 subjects, and the other of 43. All 93 subjects have logged VBR measures obtained through the use of planimetry based on the largest brain section and on the next largest section. These two measures are referred to as PLAN1 and PLAN2, respectively. The subjects in the first sample ($g = 1, N_g = 50$) have repeat measurements for both the largest and second-largest brain sections (PLAN3 and PLAN4, respectively). The subjects in the second sample ($g = 2, N_2 = 43$) do not have these repeated measurements and these have been replaced in the dispersion matrix by pseudovariables PV1 and PV2. Note that the variances of PV1 and PV2 are both unity and their

Table 5.15 Summary statistics (covariance matrix) for logged VBR data based on repeated planimetry measurements on the largest slice (PLAN1 and PLAN3) and the second largest slice (PLAN2 and PLAN4)

(a) Sample 1 (50 cases)

	PLAN1	PLAN2	PLAN3	PLAN4
PLAN1	0.163			
PLAN2	0.141	0.328		
PLAN3	0.139	0.149	0.192	
PLAN4	0.125	0.181	0.124	0.208

(b) Sample 2 (43 cases)
PV1 and PV2 are pseudovariables corresponding to PLAN3 and PLAN4, respectively.

	PLAN1	PLAN2	PV1	PV2
PLAN1	0.283			
PLAN2	0.160	0.245		
PV1	0	0	1	
PV2	0	0	0	1

covariances with each other and with any other measurement are all set at zero. This is the form in which the two covariance matrices are presented in a program such as LISREL.

Now consider a possible measurement model. We will assume that PLAN1 and PLAN3 are a pair of parallel measures. We will also make the same assumption concerning PLAN2 and PLAN4. If a single common factor is sufficient to account for the correlations between the four measurements then the data can be represented by (ignoring group membership)

$$X_{ik} = \beta_k \mu_i + e_{ik} \tag{5.59}$$

where X_{ik} is the logged VBR measure for subject i using the kth method. (Here $k = 1$ for PLAN1, 2 for PLAN2, 3 for PLAN3 or PV1 and 4 for PLAN4 or PV2.) Let $\beta_1 = \beta_3 = 1$ and $\beta_2 = \beta_4$. Also let $\sigma_1^2 = \sigma_3^2$ and $\sigma_2^2 = \sigma_4^2$ where σ_k^2 is the variance of the measurement errors for the kth measure. The parameters to be estimated are β_2, σ_1^2, σ_2^2 and σ_μ^2. This is the full description of the model for sample 1. For the second sample (with measurements PLAN1 and PLAN2 only) we have to introduce the following constraints for the parameters concerning pseudovalues: $\sigma_3^2 = \sigma_4^2 = 1$ and $\beta_3 = \beta_4 = 0$. The free parameters for the second sample (β_2, σ_1^2, σ_2^2 and σ_μ^2) are constrained to be equal to those of the first sample.

The results of fitting this CFA model are shown in Table 5.16. Note that the computer cannot distinguish between real variables and pseudovariables.

Table 5.16 Maximum likelihood estimates for CFA model of CAT scan data in Table 5.15

$\hat{\beta}_2 = 0.906$ (s.e. 0.094)	$\hat{\sigma}_1^2 = 0.046$ (s.e. 0.009)
$\hat{\sigma}_\mu^2 = 0.172$ (s.e. 0.031)	$\hat{\sigma}_2^2 = 0.127$ (s.e. 0.018)
Chi-square = 18.60 with 9 d.f.	

Accordingly it will give the incorrect number of degrees of freedom (i.e. 16) for the log-likelihood ratio statistic. The correct number of degrees of freedom is 9 (16 − 7 or 13 − 4). It appears that the model does not provide a particularly good fit to the data. It will be left to the reader to find a better one (see Exercise 5.8).

The data shown in Table 5.17 comprise estimates of the duration of episodes of spike and wave formation recorded by electrodes attached to the scalp of an epileptic child. MON1 and MON2 are measurements produced from visual examination of a trace produced using a Monolog recorder (Micromed Limited), and EEG1 and EEG2 are measurements based on visual examination of EEG traces.

Table 5.17 Measurements of the duration* of spike and wave formation (Child A)

EVENT	MON1	EEG1[†]	MON2	EEG2[†]	EVENT	MON1	EEG1[†]	MON2	EEG2[†]
1	25	25	25	25	19	5	20	10	20
2	110	105	110	105	20	45	30	50	35
3	15	0	15	10	21	30	30	30	25
4	150	160	155	155	22	35	25	35	25
5	50	40	50	40	23	80	70	80	70
6	15	15	15	15	24	130	150	130	180
7	100	90	100	100	25	45	0	45	0
8	50	20	50	30	26	20	15	20	20
9	55	50	55	55	27	75	70	75	70
10	80	80	85	80	28	55	50	55	50
11	100	100	100	100	29	20	20	20	20
12	45	40	45	40	30	135	130	130	130
13	75	50	75	50	31	5	0	10	10
14	60	60	65	60	32	20	20	20	20
15	15	0	15	20	33	65	90	65	85
16	95	80	95	75	34	10	0	10	0
17	100	90	100	90	35	30	25	30	30
18	130	130	130	130	36	40	40	40	40
					37	20	20	20	30
					Mean	57.7	52.4	58.4	55.1
					s.d.	40.5	44.1	40.3	44.3

* in units of $\frac{1}{10}$ s.
[†] A duration of 0 indicates that an event has been detected by the Monolog recorder but not from visual examination of the EEG.

The visual examinations were carried out by two trained EEG technicians – one providing MON1 and EEG1 and the other providing MON2 and EEG2. Monolog traces and EEG traces were both rated blind as part of a larger investigation of the reliability and validity of the Monolog recorder. A typical chart recording of an individual episode of spike and wave formation is illustrated in Figure 5.2.

Table 5.18 provides summaries of *two* periods of EEG Monolog recording for a second epileptic child. It is these data that will be analysed here. We will start by assuming that MON1 and MON2 comprise a pair of congeneric measures and that EEG1 and EEG2 also comprise a pair of congeneric measurements. It will also be assumed that the errors in the technicians' ratings of

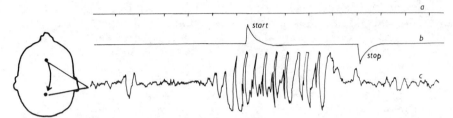

Fig. 5.2   Example of a segment of an EEG trace (c) with an accompanying Monolog signal (b). The top line (a) indicates the passage of time in seconds. The beginning of a period of spike and wave activity as recognised by the Monolog recorder is labelled 'start'. The end is labelled 'stop'.

the Monolog traces are not correlated with the corresponding errors in their assessments based on the EEG recordings. The CFA model chosen for these data will be similar to that used for the CAT scans in Section 5.5. We have

$$X_{ij1} = \beta_1 \mu_i + e_{ij1} \tag{5.60}$$

and

$$X_{ij2} = \beta_2 \eta_i + e_{ij2} \tag{5.61}$$

where X_{ijk} is the jth estimate of duration of the ith event using method k ($k = 1$ for Monolog and $k = 2$ for EEG). The variances of the common factors (i.e. σ_μ^2 and σ_η^2) will both be constrained to unity. Because of the relative simplicity of the task it will be assumed that the variance of the errors for MON1 and MON2 are the same ($\sigma_1^2 = \sigma_3^2$) but that the two technicians might differ in their precision in assessing the duration of events from the EEG traces. The measurement error for EEG1 will be σ_4^2. The correlation between μ_i and η_i is represented by ρ.

Table 5.18   Measurements of the duration (in units of $\frac{1}{10}$ s) of spike and wave formation (Child B)

(a) First period of recording (number of events = 31)

Correlation/covariance Matrix *

	MON1	EEG1	MON2	EEG2	Mean	s.d.
MON1	1198.62	0.944	0.999	0.979	29.097	34.621
EEG1	912.09	778.12	0.949	0.958	32.419	27.895
MON2	1203.81	921.72	1211.32	0.980	29.581	34.804
EEG2	967.31	762.44	973.03	813.91	26.226	28.529

(b) Second period of recording (number of events = 30)

Correlation/covariance matrix *

	MON1	EEG1	MON2	EEG2	Mean	s.d.
MON1	1274.40	0.963	0.998	0.972	43.500	35.699
EEG1	1199.83	1218.51	0.967	0.983	44.333	34.907
MON2	1258.86	1193.46	1249.44	0.978	43.267	35.347
EEG2	1182.59	1169.37	1178.80	1161.61	44.333	34.023

* Variances on the diagonal, correlations above the diagonal, and covariances below.

Table 5.19 Results of fitting CFA models to the two covariance matrices in Table 5.18

(a) No parameters invariant

Period 1	Period 2
$\hat{\beta}_1 = 34.696$ (s.e. 4.481)	$\hat{\beta}_1 = 35.480$ (s.e. 4.664)
$\hat{\beta}_2 = 28.174$ (s.e. 3.674)	$\hat{\beta}_2 = 34.053$ (s.e. 4.495)
$\hat{\sigma}_1^2 = 1.160$ (s.e. 0.030)	$\hat{\sigma}_1^2 = 3.060$ (s.e. 0.804)
$\hat{\sigma}_2^2 = 58.280$ (s.e. 17.218)	$\hat{\sigma}_2^2 = 34.163$ (s.e. 12.570)
$\hat{\sigma}_4^2 = 8.882$ (s.e. 8.677)	$\hat{\sigma}_4^2 = 7.217$ (s.e. 9.005)
$\hat{\rho} = 0.985$ (s.e. 0.007)	$\hat{\rho} = 0.980$ (s.e. 0.009)

Chi-square = 12.97 with 8 d.f.

(b) ρ invariant from Period 1 to Period 2

Period 1	Period 2
$\hat{\beta}_1 = 33.833$ (s.e. 4.020)	$\hat{\beta}_1 = 36.468$ (s.e. 4.397)
$\hat{\beta}_2 = 27.534$ (s.e. 3.326)	$\hat{\beta}_2 = 35.001$ (s.e. 4.249)
$\hat{\sigma}_1^2 = 1.159$ (s.e. 0.299)	$\hat{\sigma}_1^2 = 3.065$ (s.e. 0.805)
$\hat{\sigma}_2^2 = 59.195$ (s.e. 17.440)	$\hat{\sigma}_2^2 = 33.589$ (s.e. 12.257)
$\hat{\sigma}_4^2 = 7.150$ (s.e. 8.087)	$\hat{\sigma}_4^2 = 8.129$ (s.e. 8.670)

$\hat{\rho} = 0.982$ (s.e. 0.006)

Chi-square = 13.23 with 9 d.f.

(c) ρ and measurement error variances invariant from Period 1 to Period 2

Period 1	Period 2
$\hat{\beta}_1 = 33.744$ (s.e. 3.931)	$\hat{\beta}_1 = 36.578$ (s.e. 4.316)
$\hat{\beta}_2 = 27.417$ (s.e. 3.231)	$\hat{\beta}_2 = 35.075$ (s.e. 4.159)

$\hat{\sigma}_1^2 = 2.094$ (s.e. 0.385)

$\hat{\sigma}_2^2 = 46.737$ (s.e. 10.541)

$\hat{\sigma}_4^2 = 7.602$ (s.e. 6.221)

$\hat{\rho} = 0.982$ (s.e. 0.006)

Chi-square = 21.68 with 12 d.f.

The first model to be fitted constrains none of the free parameters to be invariant across the two periods of recording. The second constrains ρ to be invariant, and the third in addition constrains the error variances to be equal. The results are shown in Table 5.19. Models (a) and (b) appear to fit the data well; model (c) is less satisfactory. It will be left to the reader (Exercise 5.9) to further explore these data.

5.8 Analysis of data with missing values

Despite the very common occurrence of data sets that are incomplete in some way, the problem of fitting a measurement model (here a CFA model) to data sets containing missing values is usually completely ignored in text books on factor analysis. Traditionally there are two common approaches to the problem, neither of which is very satisfactory. The first is to only carry out the factor analysis on data from subjects who have complete records (i.e. no missing data). This is a reasonable procedure to use as long as the missing observations are fairly rare, but is not a very effective use of the data if most of the subjects have missing values for at least one of the required measurements. The second approach is to calculate covariances using subjects who have complete data for each pair of variables under consideration. Brown (1983) refers to the first method as the *deletion method* and following Heiberger (1977) he refers to the second as the *pair-wise present method*. Much more satisfactory approaches are discussed by Finkbeiner (1979), Brown (1983), Rubin and Thayer (1982) and Little and Rubin (1987). In this section we will concentrate on the method suggested by Muthén *et al.* (1987), and by Werts *et al.* (1979).

A satisfactory analysis is clearly dependent on making correct assumptions concerning the mechanism that has generated the missing values. In the present discussion it will be assumed that the missing data are *missing completely at random* (Little and Rubin, 1987). For many data sets this is clearly unrealistic but where data are missing *by design* it should not be. More complicated situations are discussed by Muthén *et al.* (1987). The reader is also referred to Lee (1986).

A perceptive reader might have noticed that the data presented in Table 5.15 might be regarded as coming from a single sample from which 43 cases have missing data. This suggests that the same method of analysis as that discussed in the previous section can be used for missing data problems (Jöreskog, 1971; Werts *et al.*, 1979). The adaptation of the method to missing data problems is developed in considerable detail by Muthén *et al.* (1987). Here we will illustrate its use through two relatively straightforward examples.

Consider a hypothetical set of measurements made in an analytical chemistry laboratory. The first two, *A* and *B*, do not involve destruction of the material under test and can be made on all specimens of material. The second two, *C* and *D*, both involve destruction of the specimen. A particular specimen can yield either *C* or *D*, but not both. A simple design for a reliability study might involve getting measurements *A* and *B* on all specimens and then randomly allocating specimens to be processed to give either *C* or *D*. An analogous situation might arise in the case of a psychiatric assessment where, in this situation, *A* and *B* are the results obtained from two fairly dissimilar psychiatric screening questionnaires, but *C* and *D* are results obtained through the use of two closely related and time-consuming psychiatric interviews. Here for reasons of cost as well as potential memory effects it might be thought impossible to expose subjects to both of these interviews.

The above example leads to two subsamples of observations, both of which contain missing data. One subsample has measurements *A*, *B* and *C*; the other has *A*, *B* and *D*. In order to analyse these measurements using the methods of Section 5.7 one might introduce a pseudovariable into each of the two subsamples. PV1 could replace *D* in the first sub-sample and PV2 could replace *C* on the second. The analysis would then simply involve constraining relevant free parameters to be invariant across the two groups. However, the use of pseudovariables is unnecessary in this example since there are an equal number of measurements per group.

Table 5.20 contains the results of a computer-simulated BIBD reliability study with 50 subjects per group. Each group provides three measurements out

Table 5.20 Results of a simulated BIBD reliability study with 50 subjects per group

Group 1

		Covariance matrix			Mean
		A	B	C	
	A	13.559			2.311
	B	19.431	34.733		3.625
	C	31.566	50.173	89.000	4.227

Group 2

		Covariance matrix			Mean
		A	C	D	
	A	11.550			1.949
	C	33.347	113.719		4.112
	D	11.322	36.007	13.719	1.719

Group 3

		Covariance matrix			Mean
		A	B	D	
	A	10.659			2.132
	B	18.439	38.804		2.954
	D	9.509	17.952	10.456	1.830

Group 4

		Covariance matrix			Mean
		B	C	D	
	B	24.016			2.355
	C	29.342	56.632		2.763
	D	9.343	13.403	5.088	1.843

of a possible four (*A*, *B*, *C* and *D*). The example described in the previous paragraph could be illustrated by analysis of Group 1 and Group 3 *alone* (i.e. ignoring Groups 2 and 4). Here, however, we will consider the analysis of the whole set of data.

The CFA model fitted to each of the four groups will have to be fitted to the appropriate augmented moments matrix (rather than the covariance matrix as in the previous section). This is because it is necessary to constrain the intercepts/means of the four measurements to be invariant across all groups. If the data are missing completely at random, as they are in this example, then the elements of the expected moments matrix should be invariant across groups. Considering measurement *A*, for example, its variance should be constrained to be equal for Groups 1, 2 and 3 (it does not occur in 4). Similarly the covariance between *A* and *B* should be invariant across Groups 1 and 3. The covariance between *A* and *C* should be invariant across Groups 1 and 2. And so on for all the four measurements and their covariances. A CFA model in which one simply constrains the elements of the augmented moments matrix across the four groups to be invariant is the appropriate null hypothesis compared with which various alternative hypotheses (actual measurement models) might be tested. A completely unrestricted model is only appropriate when one is comparing several populations. To re-iterate this point, in the case of a single population with several alternative patterns of missing data (i.e. subgroups) the starting point of the analysis is a model in which one simply restricts the means, variances and covariances to be invariant from one subgroup to another. This starting point also provides a test of whether the data are, indeed, missing completely at random (Muthén *et al.*, 1987). Here one is simply testing the hypothesis that the sample moment matrices are consistent with the proposition that the subjects (in the present example) have been allocated to the four groups at random. It will be left for the reader to test this hypothesis (Exercise 5.10).

The data in Table 5.20 were generated using a congeneric tests model for the four measurements. The single common factor was generated as a standard normal deviate. The error terms were also normal deviates with zero mean and with variances given in Table 5.21 (column (b)). Values of the other parameters of model are also given in column (b) of Table 5.21. Maximum likelihood estimates of these parameters, together with their standard errors, are also shown in Table 5.21 (columns (c) and (d) respectively). Details of the required model are described below.

Table 5.21 Measurement model for the BIBD data in Table 5.20 $(X_k = \alpha_k + \beta_k \mu + e)$

(a) Parameter	(b) True value	(c) Estimate	(d) Standard error
β_A	3	3.054	0.180
β_B	6	5.550	0.328
β_C	9	8.748	0.520
β_D	3	2.943	0.176
α_A	2	2.055	0.234
α_B	3	2.953	0.427
α_C	4	3.711	0.674
α_D	2	1.880	0.228
σ_A^2	1	1.140	0.210
σ_B^2	4	3.885	0.703
σ_C^2	9	9.963	1.774
σ_D^2	1	1.198	0.207

Consider group g ($g = 1, 2, 3$ or 4) and measurement k within this group ($k = 1, 2$ or 3). The appropriate measurement model (ignoring subscripts for subjects) is

$$X_{kg} = \alpha_{kg} + \beta_{kg}\mu + e_{kg} \qquad (5.62)$$

For Group 1 $k = 1$ for measurement A, $k = 2$ for measurement B and $k = 3$ for measurement C. For Group 2 $k = 1$ for measurement A, $k = 2$ for C and $k = 3$ for D. And so on. The variance of μ is 1 and the variance of e_{kg} is σ_{kg}^2. Fitting a fully constrained congeneric tests model to the four groups in Table 5.20 is equivalent to fitting a simple congeneric tests model of the form given in Equation (5.62) to each of the four groups with the following constraints:

$$\alpha_{11} = \alpha_{12} = \alpha_{13}$$
$$\alpha_{21} = \alpha_{23} = \alpha_{14}$$
$$\alpha_{31} = \alpha_{22} = \alpha_{24}$$
$$\alpha_{32} = \alpha_{33} = \alpha_{34}$$
(5.63)

$$\beta_{11} = \beta_{12} = \beta_{13}$$
$$\beta_{21} = \beta_{23} = \beta_{14}$$
$$\beta_{31} = \beta_{22} = \beta_{24}$$
$$\beta_{32} = \beta_{33} = \beta_{34}$$
(5.64)

$$\sigma_{11}^2 = \sigma_{12}^2 = \sigma_{13}^2$$
$$\sigma_{21}^2 = \sigma_{23}^2 = \sigma_{14}^2$$
$$\sigma_{31}^2 = \sigma_{22}^2 = \sigma_{24}^2$$
$$\sigma_{32}^2 = \sigma_{33}^2 = \sigma_{34}^2$$
(5.65)

The imposition of all of these constraints in LISREL is a little tedious and requires some care, but otherwise is quite straightforward.

5.9 Longitudinal factor analysis and structural equation models

The purpose of this section is to introduce methods of analysis appropriate to the study of reliability and stability in what sociologists call *panel data* (Markus, 1979). Panel data are typically thought of as information or measurements obtained from a sample of respondents (a *panel*) at two or more points in time. Each time point is referred to as a *wave* of the panel study. The number of measurements made at each time point is usually referred to as the number of *indicators*. The data described in Section 3.4 in the context of the Wiley and Wiley model of stability are an example of that obtained from a three-wave single indicator panel study. Table 5.5 might also be thought of as a description of a three-wave single indicator study. A detailed discussion of single-indicator models is provided by Jagodzinski and Kühnel (1987). Here the discussion will concentrate on the use of multiple indicators at each of the waves of the panel study. The latter are covered in Wheaton *et al.* (1977), Jagodzinski *et al.* (1987), and by Raffalovich and Bohrnstedt (1987).

First, consider a simple two-wave two-indicator panel study. At each of two distinct times two measurements are made on a panel of I subjects. Each subject thus provides a vector of four measurements from which a covariance or moments matrix can be calculated. The two measures, for example, might be scores on two closely related reading tests administered when the panel of children are 10 years old and again when they are aged 14. Let X_{ijk} represent the reading score for subject i at time j ($j = 1, 2$) using test k ($k = 1, 2$). Assuming that the test scores are measured as deviations from their respective means, a suitable measurement model might have the following form:

$$X_{i11} = \mu_{i1} + e_{i11}$$

$$X_{i12} = \lambda\mu_{i1} + e_{i12}$$

$$X_{i21} = \mu_{i2} + e_{i21} \tag{5.66}$$

$$X_{i22} = \lambda\mu_{i2} + e_{i22}$$

This is a variant of the by now familiar two-factor model in which (from this example) μ_{i1} and μ_{i2} represent the true reading ability at ages 10 and 14, respectively. The true scores of μ_{i1} and μ_{i2} are both constrained to have zero mean with variances represented by ω_1 and ω_2, respectively. The true reading abilities at the two ages are assumed to be correlated, and this correlation is represented by ρ. If the variances of e_{i11} and e_{i21} are constrained to be equal (σ_1^2) and similarly if the variances of e_{i12} and e_{i22} are constrained to be equal (σ_2^2) then the expected covariance matrix for the four measurements is that given in Table 5.22(a).

Table 5.22 Expected covariance matrix for the two-wave two-indicator panel model described by Equation (5.66)

(a) Factor analysis model

	X_{11}	X_{12}	X_{21}	X_{22}
X_{11}	$\omega_1 + \sigma_1^2$			
X_{12}	$\lambda\omega_1$	$\lambda^2\omega_1 + \sigma_2^2$		
X_{21}	$\rho\omega_1^{1/2}\omega_2^{1/2}$	$\lambda\rho\omega_1^{1/2}\omega_2^{1/2}$	$\omega_2 + \sigma_1^2$	
X_{22}	$\lambda\rho\omega_1^{1/2}\omega_2^{1/2}$	$\lambda^2\rho\omega_1^{1/2}\omega_2^{1/2}$	$\lambda\omega_2$	$\lambda^2\omega_2 + \sigma_2^2$

(b) Structural equation model

	X_{11}	X_{12}	X_{21}	X_{22}
X_{11}	$\omega_1 + \sigma_1^2$			
X_{12}	$\lambda\omega_1$	$\lambda^2\omega_1 + \sigma_2^2$		
X_{21}	$\beta\omega_1$	$\lambda\beta\omega_1$	$\beta^2\omega_1 + \psi + \sigma_1^2$	
X_{22}	$\lambda\beta\omega_1$	$\lambda^2\beta\omega_1$	$\lambda(\beta^2\omega_1 + \psi)$	$\lambda^2(\beta^2\omega_1 + \psi) + \sigma_2^2$

If, instead of simply accepting the measurement model in Equation (5.66) together with the descriptive statement that $\text{Cov}(\mu_1, \mu_2) = \rho$, we were to suggest a *causal* link between μ_{i1} and μ_{i2} such as

$$\mu_{i1} = \theta_{i1}$$

$$\mu_{i2} = \beta\mu_{i1} + \theta_{i2} \tag{5.67}$$

for example, then the expected covariance matrix would be that given in Table 5.22(b). In the derivation of this matrix it has been assumed that $\text{Cov}(e_{i1}, \theta_{i2}) = 0$, $E(e_{i2}) = 0$ and that $\text{Var}(\theta_{i2}) = \psi$. It is also assumed that $\text{Cov}(e_{i2}, e_{ijk}) = 0$ for all j and k. Equation (5.67) is a *structural equation model* similar to that proposed in the Wiley stability model (see Section 3.4).

The parameters in either form of the model for this two-wave two-indicator model are identified and can be estimated directly using a general-purpose program such as LISREL. If the factor analysis model is fitted, however, it is easy to derive estimates of β and ψ from those of ω_1, ω_2 and ρ. These are

$$\hat{\beta} = \frac{\hat{\rho}\omega_2^{1/2}}{\hat{\omega}_1^{1/2}} \tag{5.68}$$

and

$$\hat{\psi} = \hat{\omega}_2(1 - \hat{\rho}^2) \tag{5.69}$$

These are obtained by simply equating the corresponding elements of the matrices in Table 5.22(a) and Table 5.22(b). Note that Equations (5.68) and (5.69) are analogous to relationships between the regression coefficient, product-moment correlation and error variance in the simple case of bivariate linear regression.

In many measurement studies the model described by Equation (5.66) will be inadequate. Quite often a model is needed which is more analogous to that used to analyse the CAT scan data in Section 5.4 (see Equation (5.34)). This is obtained by the introduction of item specific factors (or errors) uncorrelated with the common factors and with specific factors associated with other items. The item specific factors are usually assumed to have a zero mean. For a three-indicator two-wave model the more realistic measurement might have the following form:

$$\begin{aligned}
X_{i11} &= \mu_{i1} + d_{i1} + e_{i11} \\
X_{i12} &= \lambda_1 \mu_{i1} + d_{i2} + e_{i12} \\
X_{i13} &= \lambda_2 \mu_{i1} + d_{i3} + e_{i13} \\
X_{i21} &= \mu_{i2} + \lambda_3 d_{i1} + e_{i21} \\
X_{i22} &= \lambda_1 \mu_{i2} + \lambda_4 d_{i2} + e_{i22} \\
X_{i23} &= \lambda_2 \mu_{i2} + \lambda_5 d_{i3} + e_{i23}
\end{aligned} \tag{5.70}$$

The term d_{ik} is the component of the ith subject's measurement that is both specific to item k and is common to first and second waves of measurement. If it is assumed that the correlation between μ_{i1} and μ_{i2} is ρ, but that the mutual correlations between ds and es are all zero, then the expected covariance matrix for this model is given in Table 5.23. Here the variances of d_{i1}, d_{i2} and d_{i3} are represented by $\sigma_{d_1}^2$, $\sigma_{d_2}^2$ and $\sigma_{d_3}^2$, respectively. Similarly the variance of the e_{ijk} is represented by $\sigma_{e_k}^2$. As before, let $\text{Var}(\mu_{i1}) = \omega_1$ and $\text{Var}(\mu_{i2}) = \omega_2$. Inspection of Table 5.23 will indicate that $\sigma_{e_1}^2$, $\sigma_{e_2}^2$, $\sigma_{e_3}^2$, λ_1, λ_2, ω_1 and ω_2 are identified, but that λ_3, λ_4, λ_5 and $\sigma_{d_1}^2$, $\sigma_{d_2}^2$ and $\sigma_{d_3}^2$ are not. The three products $\lambda_3\sigma_{d_1}^2$, $\lambda_4\sigma_{d_2}^2$ and $\lambda_5\sigma_{d_3}^2$ are, however, identified. The problem is solved by letting $\lambda_3 = \lambda_4 = \lambda_5 = 1$. Even more general *longitudinal factor analysis* models are described by Raffalovich and Bohrnstedt (1987). In general, for these models to be identified without imposing equality constraints, one needs at least three indicators for each of at least three waves of measurement. An interesting commentary on the above paper is provided by Zeller (1987).

Table 5.23 Expected covariance matrix for the two-wave three-indicator panel model described in Equation (5.70)

	X_{11}	X_{12}	X_{13}	X_{21}	X_{22}	X_{23}
X_{11}	$\omega_1 + \sigma_{d_1}^2 + \sigma_{e_1}^2$					
X_{12}	$\lambda_1\omega_1$	$\lambda_1^2\omega_1 + \sigma_{d_2}^2 + \sigma_{e_2}^2$				
X_{13}	$\lambda_2\omega_1$	$\lambda_1\lambda_2\omega_1$	$\lambda_2^2\omega_1 + \sigma_{d_3}^2 + \sigma_{e_3}^2$			
X_{21}	$\rho\omega_1^{1/2}\omega_2^{1/2} + \lambda_3\sigma_{d_1}^2$	$\rho\lambda_1\omega_1^{1/2}\omega_2^{1/2}$	$\rho\lambda_2\omega_1^{1/2}\omega_2^{1/2}$	$\omega_2 + \lambda_3^2\sigma_{d_1}^2 + \sigma_{e_1}^2$		
X_{22}	$\rho\lambda_1^2\omega_1^{1/2}\omega_2^{1/2}$	$\rho\lambda_1^2\omega_1^{1/2}\omega_2^{1/2} + \lambda_4\sigma_{d_2}^2$	$\rho\lambda_1\lambda_2\omega_1^{1/2}\omega_2^{1/2}$	$\lambda_1\omega_2$	$\lambda_1^2\omega_2 + \lambda_4^2\sigma_{d_2}^2 + \sigma_{e_2}^2$	
X_{23}	$\rho\lambda_2^2\omega_1^{1/2}\omega_2^{1/2}$	$\rho\lambda_1\lambda_2\omega_1^{1/2}\omega_2^{1/2}$	$\rho\lambda_2^2\omega_1^{1/2}\omega_2^{1/2} + \lambda_5\sigma_{d_3}^2$	$\lambda_2\omega_2$	$\lambda_1\lambda_2\omega_2$	$\lambda_2^2\omega_2 + \lambda_5^2\sigma_{d_3}^2 + \sigma_{e_3}^2$

Returning to Equation (5.70), here are two possible estimates of reliability possible. One is the communality of the measurement. For X_{i11} this is

$$R_c = \frac{\omega_1 + \sigma^2_{d_1}}{\omega_1 + \sigma^2_{d_1} + \sigma^2_{e_1}} \qquad (5.71)$$

The other is

$$R_e = \frac{\omega_1}{\omega_1 + \sigma^2_{d_1} + \sigma^2_{e_1}} \qquad (5.72)$$

Which of these is the most appropriate is dependent on the interpretation of the item-specific factors. If they are *not* errors then R_c is the appropriate reliability measure, but if they are errors as was assumed with the CAT scan data in Section 5.5, then R_e is better.

One important characteristic of many panel studies is sample attrition. If subjects are lost from the study at random then the data should be analysed using the methods of the previous section. For more complicated models of sample attrition the reader is referred to Muthén *et al.* (1987).

5.10 Further consideration of sample size requirements

Sample size requirements were briefly discussed in Section 4.9. The approach taken here will be less formal than the previous one but will be more relevant to the analysis of covariance matrices. One simple way of getting some idea of the likely adequacy of the size of a particular sample is to look at the standard errors obtained in previous analyses (see Tables 5.9, 5.16, 5.19 and 5.21, for example). One major problem that has repeatedly caused difficulties in fitting factor analysis and structural equation models is the occurrence of improper estimates (negative variances, for example). This can be caused by badly specified or inappropriate models, poor starting values and quite often from sampling variability. The latter is an indication of inadequate sample sizes. Some of these problems are discussed by Bentler and Chou (1987), Boomsma (1985) and Rindskopf (1984).

Consider the estimation of the reliability of a particular measuring instrument. It has been suggested earlier in this chapter that an investigation of the properties will require at least two measurements on each subject in addition to the one of interest. It is clear that the quality of the reliability estimates, as indicated, for example, by a standard error or confidence interval, will improve as the sample of subjects gets larger. But how dependent is the quality of the reliability estimate on the precision of the concomitant measurements? Presumably one should use the best instruments available and if this is not practicable or possible then the sample size should be increased to compensate for the poorer quality of the additional measurements. Another possibility is to increase the number of concomitant measurements to compensate for their lack of quality.

The above ideas will now be illustrated by a series of computer-simulated reliability studies. The results will not provide a recipe for sample size estimation, nor will they lead to the choice of the 'best' design. They will simply explore some of the possibilities. In the simulation studies it will be assumed that one is interested in investigating the reliability of one of a battery of congeneric tests. The data will be used to explore the effects of (a) changing sample size, (b) changing the reliabilities of the concomitant tests, and (c) increasing the number of concomitant tests.

The model chosen for each test has the following form (ignoring subscripts for subjects):

$$X_k = 3T + e_k \qquad (5.73)$$

where X_k is the measurement obtained using the kth test and e_k is the corresponding measurement error. The underlying latent variable, T, is generated to be distributed as a standard normal deviate (i.e. mean zero and unit variance). The error, e_k, is also generated as a normal deviate, independent of T, with mean zero but with variance dependent on the particular test. For the test whose reliability is to be estimated data is generated with an error variance of unity so that its true reliability is always 0.90. The errors for each of the concomitant tests are generated to be independent of each other and independent of the measurement errors for the target test. For each particular choice of sample size, number of concomitant tests, and their quality 1000 data sets were generated using the computer package GLIM (see Appendix 4). Each of these 1000 data sets then yielded a reliability estimate for the target test and then the sampling distribution of the 1000 estimates was tabulated.

In the case of three congeneric tests (including the target test) the reliability of the test was calculated for each sample using Equations (5.6) and (5.12). In the case of four congeneric tests there are three possible estimates of β_1^2, that is

$$\beta_1^2 = \frac{S_{12} S_{13}}{S_{23}}$$

or

$$\frac{S_{12} S_{14}}{S_{24}} \qquad (5.74)$$

or

$$\frac{S_{13} S_{14}}{S_{34}}$$

In the present study these three estimates were simply averaged, providing a combined estimate that is not identical to the maximum likelihood estimate but is close enough to it for the present purposes. A similar approach was also used in the case of experiments involving five or more congeneric tests.

In the first three experiments (A1, A2 and A3) three congeneric tests were used, each with a true reliability of 0.90. These three were simply used to illustrate the effect of sample size. The sample sizes for experiments A1, A2 and A3 were 50, 25 and 15, respectively. In experiments B and C the second concomitant test had a reliability of 0.70 and 0.50, respectively (the other two still have true reliabilities of 0.90). In experiment D both of the concomitant tests had a true reliability of 0.70. In experiments E and F there were three and four concomitant tests, respectively, all with a reliability of 0.70. In experiments B–F the sample size was fixed at 50. A summary of the various designs is given in Table 5.24, and the results are shown in Table 5.25.

Table 5.24 Simulation of congeneric tests

| Experiment | Reliabilities | | | | | Sample size |
	R1*	R2	R3	R4	R5	
A1	0.9	0.9	0.9	—	—	50
A2	0.9	0.9	0.9	—	—	25
A3	0.9	0.9	0.9	—	—	15
B	0.9	0.9	0.7	—	—	50
C	0.9	0.9	0.5	—	—	50
D	0.9	0.7	0.7	—	—	50
E	0.9	0.7	0.7	0.7	—	50
F	0.9	0.7	0.7	0.7	0.7	50

* Test 1 is the target measurement. Tests 2 to 5 are the concomitant measurements.

The results are fairly clear-cut and do not require much explanation. Here we will simply concentrate on the number of estimates of R1 that fell below 0.80 (as an indication of lack of precision) and the number of Heywood cases (where the estimate of R1 would have been greater than or equal to 1.00). Lowering the sample size has a clear impact on both of these summary counts. Lowering the reliabilities of the concomitant tests (experiments B, C and D) also has a clear effect. The sample size would have to be significantly increased above 50 to make up for the impact of using poor quality measuring instruments.

Table 5.25 Summary of the results of the simulation experiments in Table 5.24: frequency distributions for estimates of R1

| Estimate | Experiment | | | | | | | |
	A1	A2	A3	B	C	D	E	F
R1 < 0.80	10	69	118	53	72	98	48	23
0.80 ≤ R1 < 0.81	7	16	20	10	25	30	17	11
0.81 ≤ R1 < 0.82	12	20	26	17	20	28	13	14
0.82 ≤ R1 < 0.83	19	31	17	25	27	34	20	18
0.83 ≤ R1 < 0.84	25	32	29	27	36	27	26	25
0.84 ≤ R1 < 0.85	34	30	32	33	36	32	30	37
0.85 ≤ R1 < 0.86	34	47	35	48	39	47	52	42
0.86 ≤ R1 < 0.87	57	50	29	53	45	39	36	59
0.87 ≤ R1 < 0.88	61	52	43	62	51	42	59	67
0.88 ≤ R1 < 0.89	102	53	44	74	55	56	67	72
0.89 ≤ R1 < 0.90	108	74	44	76	59	56	62	69
0.90 ≤ R1 < 0.91	118	84	59	85	59	57	76	93
0.91 ≤ R1 < 0.92	106	76	69	82	68	54	69	87
0.92 ≤ R1 < 0.93	100	62	66	74	69	51	76	86
0.93 ≤ R1 < 0.94	78	73	57	56	53	49	63	73
0.94 ≤ R1 < 0.95	59	73	77	73	49	41	57	77
0.95 ≤ R1 < 0.96	37	54	53	40	42	41	62	49
0.96 ≤ R1 < 0.97	20	44	45	27	37	38	52	37
0.97 ≤ R1 < 0.98	6	27	39	29	38	29	40	25
0.98 ≤ R1 < 0.99	5	16	32	15	21	31	18	12
0.99 ≤ R1 < 1.00	2	10	27	18	21	26	23	13
Heywood cases	0	7	39	23	78	94	32	11

Some of the loss can be made up by using additional tests, the rate of improvement presumably being comparable to that expected from an inspection of the Spearman–Brown prophesy formula (see Equation (3.24)).

5.11 Exercises

5.1 (a) Compute Cronbach's alpha coefficient from the data in Table 1.2 and estimate its standard error using both bootstrap and jackknife techniques. (b) Calculate the intra-class correlation for the two sets of measurements in Table 1.1 and use bootstrap sampling to construct a 95% confidence interval for this estimate.

5.2 Investigate the sampling characteristics of the product-moment correlation, the intra-class correlation and the alpha coefficient given in Table 5.4. Compare the standard errors obtained through the use of the jackknife with the corresponding bootstrapped standard errors.

5.3 Analyse the data given in Table 5.26 and compare the results of the analysis of the measurements in Table 5.4 (see Section 5.3).

Table 5.26 Comparison of two kitchen scales: data set 2 (weights in grammes)

Item	Scale A	Scale B	A − B	A + B
1	300	320	− 20	620
2	190	190	0	380
3	80	90	− 10	170
4	20	50	− 30	80
5	200	220	− 20	420
6	550	550	0	1100
7	400	410	− 10	810
8	610	600	10	1210
9	740	760	− 20	1500
10	1040	1080	− 40	2120
11	920	940	− 20	1860
12	1160	1180	− 20	2340
13	1330	1330	0	2660
14	1490	1510	− 20	3000
15	1620	1590	30	3210
16	1360	1330	30	2690
17	1150	1140	10	2290
18	1000	980	20	1980
19	580	550	30	1130

5.4 Fit the model described by Equation (5.4) directly to the data in Table 5.11(a) using maximum likelihood and compare your results with those given in Table 5.13. Introduce the constraint $\sigma_{d_2}^2 = 0$ and repeat the fit. Compare the parameter estimates with those obtained by use of unweighted least squares and generalised least squares as fitting criteria.

5.5 Consider four different assessments of psychiatric distress (Y1, Y2, Y3 and Y4). Y1 is obtained from the use of a semi-structured interview, The Clinical Interview Schedule (Goldberg *et al.*, 1970), and Y2 is obtained from a computerized adaptation of this schedule (Lewis *et al.*, 1988). Both Y3 and

Y4 are obtained from self-completion questionnaires: Y3 from the Hospital Anxiety and Depression Scale (Zigmond and Snaith, 1983), and Y4 from the General Health Questionnaire (Goldberg, 1972). The following matrix (provided by Dr Glyn Lewis) contains the inter-correlations between these four psychiatric measures as determined from a sample of 44 subjects.

	Y1	Y2	Y3	Y4
Y1	1			
Y2	0.860	1		
Y3	0.744	0.793	1	
Y4	0.866	0.806	0.738	1

Their standard deviations were 11.31, 6.59, 6.27 and 6.49, respectively. Fit a congeneric tests model using maximum likelihood and use the resulting parameter estimates to determine the reliabilities of the four measures. Comment on your results. Discuss potential pitfalls of the use of CFA models in the analysis of data from psychiatric questionnaires.

5.6 Consider the following covariance matrix for episode durations from a third epileptic child (subject C).

	MON1	EEG1	MON2	EEG2
MON1	18218.4			
EEG1	18087.3	18915.5		
MON2	18363.0	18173.7	18540.8	
EEG2	19063.5	19493.5	19189.8	20510.1

Fit the two-factor model described by Equations (5.60) and (5.61) to this data with the additional constraint $\sigma_2^2 = \sigma_4^2$. Investigate the influence of the starting value for ρ on the solutions.

5.7 Analyse the data in Table 5.17. Check the assumptions for any models fitted by preliminary graphical exploration of the data.

5.8 Analyse the data on CAT scans given in Table 5.15 through the use of a two-factor model. Use your results to estimate the reliabilities for the VBR measurements corresponding to the two different brain 'slices'.

5.9 Consider the data in Table 5.18. Test for (a) equality of the two covariance matrices and (b) equality of the two correlation matrices. Test for the equality of σ_2^2 and σ_4^2 as described by the model fitted in Table 5.19(b) and also in Table 5.19(c). Test whether $\hat{\rho}$ given in Table 5.19(b) is significantly different from unity. Fit any other models that you think might be useful to further explore the data in Table 5.18.

5.10 For the data in Table 5.20 test the invariance of the appropriate elements of the augmented moments matrices across (a) G1 and G3 and (b) across all four groups. Test the fit of the congeneric tests model whose estimates are given in Table 5.21.

6

Components of Variance

6.1 A simple replication study

At the very beginning of this book (Section 1.1) the concept of precision of a measuring instrument was introduced through the example of a cook making repeated weighings of a single object using kitchen scales. A slightly more complicated experiment could have involved repeated weighings of each of a random sample of objects from the cook's kitchen. Table 6.1 presents the results of an analogous reliability experiment using a simple planimeter. Each of ten map distances is measured five times, yielding a table of 50 measurements. Table 6.1 also gives the mean for each set of five measurements together with their variance. On the assumption that the precision of the planimeter is independent of the true length of the route being measured, this can be estimated from the mean of the ten error variances. The mean value is 0.413 with a standard error of 0.072. One can check whether the error variances are indeed homogeneous using one of several commonly used significance tests. British Standard 5497 recommends the use of Cochran's one-sided outlier test (see British Standards Institute, 1987 or Caulcott and Boddy, 1983). One should also be on the look-out for statistical outliers. These can usually be detected from a visual inspection of the data, but BS 5497 recommends that if the result of applying Cochran's test is significant, then this test should be followed by the use of Dixon's outlier test (see Dixon and Massey, 1969; British Standards Institute, 1987; or Caulcott and Boddy, 1983).

The above estimate of the within-route error variance could have also been obtained through the use of a one-way *analysis of variance* (ANOVA). It is assumed that the reader is familiar with this technique (see Snedecor and Cochran, 1967; or Armitage and Berry, 1987), and only a brief outline to introduce the necessary terminology will be given here. Let the jth measurement on the ith route be represented by X_{ij} ($i = 1, 2, \ldots, 10; j = 1, 2, \ldots, 5$). Note that there is no correspondence between the jth measurement of any one route with the jth measurement of another. The repeated measurements are *nested* within routes and a better representation of the measurements might be $X_{j(i)}$ to indicate the jth replication within route i. Where there is no risk of confusion, however, the simpler terminology will be used. The mean of the

Table 6.1 Measurements of map distances using a planimeter (arbitrary units)

Route	Replicate measurements					Mean	Variance
1	48,	47.5,	48,	49.5,	48	48.2	0.575
2	32,	32.5,	33.5,	31.5,	32.5	32.4	0.550
3	12.5,	14,	14,	14.5,	14.5	13.9	0.675
4	29.5,	30,	29,	29,	30	29.5	0.250
5	50,	50,	49.5,	49.5,	50.5	49.9	0.175
6	55.5,	55.5,	56,	56,	56.5	55.9	0.175
7	22,	23.5,	23,	24,	22	22.9	0.800
8	50.5,	50,	51.5,	51,	50	50.6	0.425
9	26.5,	26,	26,	25.5,	25.5	25.9	0.175
10	24,	24,	24.5,	23,	23.5	23.8	0.325
						Mean	0.413
						s.d.	0.228

five measurements for route i will be represented by $\bar{X}_i$ and the overall mean for all 50 measurements by $\bar{X}$.

In the more general situation where there are n_i measurements made on the ith subject (where $i = 1, 2, \ldots, I$ and $\sum_i n_i = N$) then

$$\bar{X}_i = \frac{1}{n_i} \sum_{j=1}^{n_i} X_{ij} \tag{6.1}$$

and

$$\bar{X} = \frac{1}{N} \sum_{i=1}^{I} \sum_{j=1}^{n_i} X_{ij} \tag{6.2}$$

$$= \frac{1}{N} \sum_{i=1}^{I} n_i \bar{X}_i \tag{6.3}$$

In the present example $I = 10$, $n_i = 5$ for all values of i and $N = 50$. The total sum of squared deviations about the overall mean is given by

$$\text{TSS} = \sum_{i=1}^{I} \sum_{j=1}^{n_i} (X_{ij} - \bar{X})^2 \tag{6.4}$$

The between-subjects (between-routes) sum of squares is given by

$$\text{BSS} = \sum_{i=1}^{I} \sum_{j=1}^{n_i} (\bar{X}_i - \bar{X})^2 \tag{6.5}$$

$$= \sum_{i=1}^{I} n_i (\bar{X}_i - \bar{X})^2 \tag{6.6}$$

Finally, the within-subjects (within-routes) sum of squares is given by

$$\text{WSS} = \sum_{i=1}^{I} \sum_{j=1}^{n_i} (X_{ij} - \bar{X}_i)^2 \tag{6.7}$$

$$= \sum_{i=1}^{I} (n_i - 1)s_i^2 \tag{6.8}$$

where s_i^2 is the within-subject error variance for subject i.

The between-subjects mean square and within-subjects mean square will be denoted by s_B^2 and s_W^2, respectively. These are calculated from

$$s_B^2 = \text{BSS}/(I - 1) \qquad (6.9)$$

and

$$s_W^2 = \text{WSS}/(N - I) \qquad (6.10)$$

An analysis of variance for the map distances given in Table 6.1 is provided in Table 6.2. It should be clear to the reader that for these data s_W^2 is equivalent

Table 6.2 Analysis of variance for the map distances given in Table 6.1

Source of variation	Degrees of freedom	Sum of squares	Mean square
Route	9	9562.000	1062.444
Error	40	16.500	0.4125

to the mean of the error variances given in Table 6.1. The standard error of s_W^2 is provided by the simple method given for Table 6.1 but it can also be estimated through the use of the bootstrap or jackknife techniques (see Section 5.2). Table 6.3 illustrates the use of the jackknife for the map distance data. The standard analysis of variance is repeated ten times, each time missing out one of the routes. Pseudovalues for the residual mean square, s_W^2, can then be calculated for each case and these, in turn, lead to the jackknife estimate of s_W^2 and its corresponding standard error. Cronbach *et al.* (1972) prefer to calculate jackknifed variance estimates and their standard errors through the use of logged variance estimates instead of the raw values as given in Table 6.3. Bootstrap methods would be equally simple to apply, but would be computationally much more expensive. The latter, however, have the advantage of leading to an easily computed confidence interval for s_W^2.

Table 6.3 Illustration of the jackknife technique in the analysis of variance of the map distances in Table 6.1

Missing route	s_W^2 estimate	s_W^2 (pseudovalue)
None	0.4125	—
1	0.4222	0.3252
2	0.4389	0.1749
3	0.4111	0.4251
4	0.3694	0.8004
5	0.4389	0.1749
6	0.4389	0.1749
7	0.4306	0.2496
8	0.3833	0.6753
9	0.3972	0.5502
10	0.3944	0.5754
Mean		0.4126
Standard error of mean		0.0722

6.2 Intra-class correlation

The intra-class correlation for pairs of measurements was introduced in Section 2.9. Here the interest will be in an index of correlation between measurements on the subject in the more general situation where there are often more than two measurements per subject. In the case of two measurements made on each subject the intra-class correlation was simply the Pearson product-moment correlation between the pairs of measurements in which each pair entered the calculation twice (i.e. as $\{X_{i1}, X_{i2}\}$ and also as $\{X_{i2}, X_{i1}\}$). Generally, if there are n_i measurements made on subject i there will be $n_i(n_i - 1)$ pairs of measurements for this one subject, each measurement being first in association with each of the other $(n_i - 1)$ measurements second. For each of the routes in Table 6.1, for example, there are 20 pairs of observations into the calculation of the product-moment correlation (200 pairs in all).

In practice, one usually computes the intra-class correlation using a different procedure to that described above. As in the previous section, let the jth measurement on subject i be represented by X_{ij} $(i = 1, 2, ..., I; j = 1, 2, ..., n_i)$. In the above correlation calculation each measurement for subject i will appear $n_i - 1$ times and so the mean of each of the variates being correlated (say X and Y) will be given by

$$\bar{X} = \bar{Y} = \frac{1}{N^*} \sum_{i=1}^{I} \left\{ (n_i - 1) \sum_{j=1}^{n_i} X_{ij} \right\} \tag{6.11}$$

where $N^* = \sum_{i=1}^{I} n_i(n_i - 1)$. Similarly,

$$\text{Var}(X) = \text{Var}(Y) = \frac{1}{(N^* - 1)} \sum_{i=1}^{I} \left\{ (k_i - 1) \sum_{j=1}^{n_i} (X_{ij} - \bar{X})^2 \right\} \tag{6.12}$$

and

$$\text{Cov}(X, Y) = \frac{1}{(N^* - 1)} \sum_{i=1}^{I} \sum_{j=1}^{n_i} \sum_{h=1}^{n_i} \{ (X_{ij} - \bar{X})(X_{ih} - \bar{X}) \} \tag{6.13}$$

where the latter only applies to pairs of measurements for which $j \neq h$. The required Pearson product-moment correlation coefficient is then given by

$$r_i = \frac{\sum_{i=1}^{I} \sum_{j=1}^{n_i} \sum_{h=1, h \neq j}^{n_i} (X_{ij} - \bar{X})(X_{ih} - \bar{X})}{\sum_{i=1}^{I} (n_i - 1) \sum_{j=1}^{n_i} (X_{ij} - \bar{X})^2} \tag{6.14}$$

But

$$\sum_{i=1}^{I} \sum_{j=1}^{n_i} \sum_{\substack{h=1 \\ h \neq j}}^{n_i} (X_{ij} - \bar{X})(X_{ih} - \bar{X}) = \sum_{i=1}^{I} \sum_{j=1}^{n_i} \sum_{h=1}^{n_i} (X_{ij} - \bar{X})(X_{ih} - \bar{X})$$

$$- \sum_{i=1}^{I} \sum_{j=1}^{n_i} (X_{ij} - \bar{X})^2 \tag{6.15}$$

where in this case the summation over j and h now extends over all possible pairs, including $h = j$. Furthermore,

$$\sum_{i=1}^{I} \sum_{j=1}^{n_i} \sum_{h=1}^{n_i} (X_{ij} - \bar{X})(X_{ih} - \bar{X}) = \sum_{i=1}^{I} n_i^2 (\bar{X}_i - \bar{X})^2 \tag{6.16}$$

where, as before, $\bar{X}_i$ is the mean value for the measurements on the ith subject. Therefore

$$r_i = \frac{\sum_{i=1}^{I} n_i^2 (\bar{X}_i - \bar{X})^2 - \sum_{i=1}^{I} \sum_{j=1}^{n_i} (X_{ij} - \bar{X})^2}{\sum_{i=1}^{I} (n_i - 1) \sum_{j=1}^{n_i} (X_{ij} - \bar{X})^2} \tag{6.17}$$

In the case where $n_i = n$ for all subjects, then

$$r_i = \frac{n^2 \sum_{i=1}^{I} (\bar{X}_i - \bar{X})^2 - \sum_{i=1}^{I} \sum_{j=1}^{n} (X_{ij} - \bar{X})^2}{(n - 1) \sum_{i=1}^{I} \sum_{j=1}^{n} (X_{ij} - \bar{X})^2} \tag{6.18}$$

Substituting in this expression the various terms obtained from an analysis of variance, it follows that

$$r_i = \frac{n \cdot \text{BSS} - \text{TSS}}{(n - 1)\text{TSS}} \tag{6.19}$$

and since TSS = BSS + WSS, it also follows that

$$\begin{aligned}
r_i &= \frac{n \cdot \text{BSS} - (\text{BSS} + \text{WSS})}{(n - 1)(\text{BSS} + \text{WSS})} \\
&= \frac{(n - 1)(I - 1)s_B^2 - I(n - 1)s_W^2}{(n - 1)[(I - 1)s_B^2 + I(n - 1)s_W^2]} \\
&= \frac{(I - 1)s_B^2 - Is_W^2}{(I - 1)s_B^2 + I(n - 1)s_W^2}
\end{aligned} \tag{6.20}$$

Further details can be found in Kendall (1943) or in Snedecor and Cochran (1967). For the data in Table 6.1, formula (6.20) gives an intra-class correlation of 0.998.

6.3   Expected mean squares

Consider the specification of a measurement model for the map distances in Table 6.1. Remembering the nested nature of the measurement errors, one possibility is

$$X_{j(i)} = \mu + \tau_i + e_{j(i)} \tag{6.21}$$

or, equivalently, as long as there is no confusion:

$$X_{ij} = \mu + \tau_i + e_{ij} \tag{6.22}$$

This is equivalent to the model described by Equation (1.1) except that, in the present case, the τ_i are departures from an overall mean μ. As in Equations (1.2) and (1.3) the expected values of the measurements are given by

$$\underset{j}{E}(X_{ij}) = \mu + \tau_i \tag{6.23}$$

and

$$\underset{i,j}{E}(X_{ij}) = \mu \tag{6.24}$$

The overall mean, μ, is assumed to be fixed; τ_i is a random variable with mean zero and variance σ_τ^2. Similarly, e_{ij} is another random variable with mean zero and variance σ_e^2. As in previous measurement models, it is assumed that errors are uncorrelated with each other and with the other components of the model. The τ_i are also assumed to be uncorrelated. From Equation (6.22) and these assumptions it follows that

$$\text{Var}(X_{ij}) = \sigma_\tau^2 + \sigma_e^2 \tag{6.25}$$

$$\text{Cov}(X_{ij}, X_{ih}) = \sigma_\tau^2 \quad (h \neq j) \tag{6.26}$$

and

$$\text{Cov}(X_{ij}, X_{kj}) = 0 \quad (i \neq k) \tag{6.27}$$

Returning to the analysis of variance described in Section 6.1 and letting $n_i = n$ for all i, from Equations (6.6) and (6.9) it follows that

$$E(s_B^2) = \frac{n}{(I-1)} E\left[\sum_{i=1}^{I} (\bar{X}_i - \bar{X})^2 \right] \tag{6.28}$$

$$= \frac{nI}{(I-1)} E(\bar{X}_i - \bar{X})^2$$

$$= \frac{nI}{(I-1)} [E(\tau_i - \bar{\tau})^2 + E(\bar{e}_i - \bar{e})^2]$$

where $\bar{\tau}$ is the mean of the τ_i, $\bar{e}_i$ is the mean of the errors for subject i and $\bar{e}$ is the mean of all of the measurement errors. Therefore

$$E(s_B^2) = \frac{nI}{(I-1)} \left[\sigma_\tau^2 - \frac{\sigma_\tau^2}{I} + \frac{\sigma_e^2}{n} - \frac{\sigma_e^2}{nI} \right]$$

$$= n\sigma_\tau^2 + \sigma_e^2 \tag{6.29}$$

Similarly from Equations (6.7) and (6.10)

$$E(s_W^2) = \frac{I}{N-I} E\left[\sum_{j=1}^{n} (X_{ij} - \bar{X}_i)^2 \right] \tag{6.30}$$

$$= \frac{nI}{N-I} E(e_{ij} - \bar{e}_i)^2$$

$$= \frac{nI}{nI-I} \left[\sigma_e^2 - \frac{\sigma_e^2}{n} \right]$$

$$= \frac{nI}{I(n-1)} \left[\frac{n-1}{n} \right] \sigma_e^2$$

$$= \sigma_e^2 \tag{6.31}$$

The within-subjects mean square is therefore an unbiased estimater of the variance of the measurement errors. Similarly, the between-subjects variance, σ_τ^2, can be estimated from

$$\hat{\sigma}_\tau^2 = \frac{s_B^2 - s_W^2}{n} \tag{6.32}$$

Finally, returning to the measurement model described by Equation (6.22), one can define the intra-class correlation by

$$\rho_i = \frac{\text{Cov}(X_{ij}, X_{ih})}{\sqrt{\text{Var}(X_{ij})}\sqrt{\text{Var}(X_{ih})}} \qquad (j \neq h) \tag{6.33}$$

$$= \frac{\text{Cov}(X_{ij}, X_{ih})}{\sqrt{\text{Var}(X_{ij})}} \tag{6.34}$$

$$= \frac{\sigma_\tau^2}{\sigma_\tau^2 + \sigma_e^2} \tag{6.35}$$

Equation (6.35) is also a definition of the measurements' reliability (see Section 3.1). One can substitute estimates of σ_τ^2 and σ_e^2 into Equation (6.35) to obtain an estimate of the instrument's reliability as

$$\hat{\rho}_i = \frac{(1/n)[s_B^2 - s_W^2]}{(1/n)[s_B^2 - s_W^2] + s_W^2}$$

$$= \frac{s_B^2 - s_W^2}{s_B^2 + (n - 1)s_W^2} \tag{6.36}$$

For the data in Table 6.1, $\hat{\rho}_i = 0.998$. Note that Equations (6.20) and (6.36) are not equivalent but, for most practical purposes, the difference is trivial.

6.4   Maximum likelihood estimators

Equations (6.25)–(6.27) define an $N \times N$ dispersion matrix $\mathbf{V}$. If it is assumed that both the τ_i and e_{ij} are normally distributed with zero means then the likelihood for the N measurements is given by

$$L = (2\pi)^{-1/2nI}|\mathbf{V}|^{-1/2} \exp\{-\tfrac{1}{2}(\mathbf{X} - \boldsymbol{\mu})'\mathbf{V}^{-1}(\mathbf{X} - \boldsymbol{\mu})\} \tag{6.37}$$

where the elements of $\mathbf{X}$ are the measurements, X_{ij}, arranged as an $N \times 1$ column vector and $\boldsymbol{\mu}$ is an $N \times 1$ of μs (all identical). Equating to zero the differentials of $\log L$ with respect to μ, σ_τ^2 and σ_e^2 provides the following solutions:

$$\hat{\mu} = \bar{X} \tag{6.38}$$

$$\hat{\sigma}_e^2 = s_W^2 \tag{6.39}$$

$$\hat{\sigma}_\tau^2 = \left(\frac{I - 1}{I} s_B^2 - s_W^2\right)\bigg/ n \tag{6.40}$$

If $[(I - 1)/I]s_B^2 - s_W^2$ is negative, then the maximum likelihood (ML) estimator of σ_τ^2 is zero and that of σ_e^2 is TSS/nI. For details the reader is referred to Searle (1971) and Searle (1987). Note that the ML estimator of σ_τ^2 is not the same as that given by Equation (6.32) and is therefore biased. Substituting the solutions in (6.39) and (6.40) into Equation (6.35) gives

$$\hat{\rho}_i = \frac{\{[(I - 1)/I]s_B^2 - s_W^2\}/n}{\{[(I - 1)/I]s_B^2 - s_W^2\}/n + s_W^2} \tag{6.41}$$

$$= \frac{(I - 1)s_B^2 - Is_W^2}{(I - 1)s_B^2 + (n - 1)Is_W^2} \tag{6.42}$$

This is identical to Equation (6.20).

A variant of ML estimation is the use of *restricted* or *residual maximum likelihood* (REML) estimators. These are described in Searle (1987), D. L. Robinson (1987) and Harville (1977). The general techniques for REML estimation were developed by Patterson and Thompson (1971; 1975). An early example of its use for the analysis of the behaviour of measuring instruments is provided by Russell and Bradley (1958).

Details of REML estimation are beyond the scope of this text but essentially the method involves maximising the joint likelihood of a set of *orthogonal contrasts* of the measurements with all of the contrasts having zero expectation (*error contrasts*). In the case of balanced data (when $n_i = n$ for all i in the one-way ANOVA models, for example) REML estimators are identical to those obtained by equating mean squares with their expected values. One exception is when the estimates obtained using expected values are negative. REML estimators are constrained to be positive or zero. Readers who are unfamiliar with the idea of *linear contrasts* and also *orthogonality* are referred to Hand and Taylor (1987) or Searle (1987) for a description of their properties. In the case of the model in (6.72) maximising the likelihood of a set of error contrasts is equivalent to maximising that part of the full likelihood that is invariant to the value of μ.

One advantage of the use of both ML and REML estimation is that the methods also provide large-sample estimates of the variances and covariances of the components. ML and REML estimates of σ_τ^2 and σ_e^2 for the data on map distances are provided in Table 6.4. This table also provides the covariance matrices for these estimates. The reader should note the size of the standard errors of the estimates and, in particular, the large standard error for σ_τ^2.

The covariance matrices for the components of variance estimates in Table 6.4 will be used to illustrate how the delta technique can provide approximate standard errors for reliability coefficients (see Appendix 3). From Equation

Table 6.4 Maximum likelihood estimates of variance components for map distances in Table 6.1

(a) ML estimates

σ_τ^2	191.159	(s.e. 85.526)
σ_e^2	0.4125	(s.e. 0.0949)

Covariance matrix

	σ_τ^2	σ_e^2
σ_τ^2	7314.71	−0.002
σ_e^2	−0.002	0.009

(b) REML estimates

σ_τ^2	212.406	(s.e. 100.168)
σ_e^2	0.4125	(s.e. 0.0922)

Covariance matrix

	σ_τ^2	σ_e^2
σ_τ^2	10033.67	−0.0017
σ_e^2	−0.0017	0.0085

(6.35) the partial differential with respect to σ_τ^2 is given by

$$\frac{\partial \rho_i}{\partial \sigma_t^2} = \frac{(\sigma_\tau^2 + \sigma_e^2) - \sigma_\tau^2}{(\sigma_\tau^2 + \sigma_e^2)^2} = \frac{\sigma_e^2}{(\sigma_\tau^2 + \sigma_e^2)^2} \qquad (6.43)$$

Similarly,

$$\frac{\partial \rho_i}{\partial \sigma_e^2} = \frac{-\sigma_\tau^2}{(\sigma_\tau^2 + \sigma_e^2)^2} \qquad (6.44)$$

Therefore

$$\mathrm{Var}(\rho_i) = \left[\frac{\sigma_e^2}{(\sigma_\tau^2 + \sigma_e^2)^2}\right]^2 \mathrm{Var}(\sigma_t^2) + \left[\frac{\sigma_t^2}{(\sigma_\tau^2 + \sigma_e^2)^2}\right]^2 \mathrm{Var}(\sigma_e^2)$$

$$-\frac{2\sigma_e^2 \sigma_\tau^2}{(\sigma_\tau^2 + \sigma_e^2)^4} \mathrm{Cov}(\sigma_\tau^2 \sigma_e^2)$$

$$= \frac{1}{(\sigma_\tau^2 + \sigma_e^2)^2} [(\sigma_e^2)^2 \, \mathrm{Var}(\sigma_\tau^2) + (\sigma_\tau^2)^2 \, \mathrm{Var}(\sigma_e^2) - 2\sigma_e^2 \sigma_\tau^2 \, \mathrm{Cov}(\sigma_e^2, \sigma_\tau^2)] \qquad (6.45)$$

For the ML estimates, $\hat{\rho}_i = 0.988$ and the standard error of $\hat{\rho}_i$ is 0.001.

6.5 Lack of balance (missing data)

In the last two sections the discussion was limited to the case where an equal number of measurements has been made on each individual. The data obtained in a study of this sort are referred to as being from a *balanced* design. In general, however, there are not exactly the same number of replicates per individual. For one reason or another there will be missing observations. This may be by accident or by design. In this case the data are referred to as being *unbalanced*. Variance components can be estimated using ML or REML programs as above, but in this case the REML estimates will not be the same as those obtained by equating mean squares (s_W^2 and s_B^2) with their expected values. REML estimators will, however, be less biased than their ML counterparts.

Returning to the analysis of variance described in Section 6.1 it is fairly straightforward to show that

$$E(s_B^2) = \left[\left(N - \sum_{i=1}^{I} n_1^2/N\right)\sigma_\tau^2 + (I-1)\sigma_e^2\right]\Big/(I-1) \qquad (6.46)$$

and

$$E(s_W^2) = \sigma_e^2 \qquad (6.47)$$

If s_B^2 and s_W^2 are equated with their expected values, then the following estimators can be obtained:

$$\hat{\sigma}_\tau^2 = \frac{s_B^2 - s_W^2}{(N - \sum_{i=1}^{I} n_1^2/N)(I-1)} \qquad (6.48)$$

and

$$\hat{\sigma}_e^2 = s_W^2 \qquad (6.49)$$

When $n_i = n$ for all, Equation (6.48) simplifies to Equation (6.32).

One particular example of an unbalanced design is likely to occur quite frequently. This is where there are two measurements made on a sample of k_2 individuals and only a single measurement made on a further k_1 individuals. Thompson and Anderson (1975) refer to this as a (k_1, k_2)-design. As an example, consider the CAT scan data that were analysed in Sections 5.5 and 5.7. Here there were complete sets of measurements for 50 subjects and incomplete data for a further 43. Table 6.5 presents the logged VBR measurements based on planimetry of the largest section (PLAN1 and PLAN3 of Tables 5.10 and 5.15). There are replicate measurements for 50 subjects ($k_2 = 50$) and single ones for 43 ($k_1 = 43$). Again it is quite straightforward to obtain ML and REML estimates of the variance components for data such as these, but in this case a usual analysis of variance table is not appropriate. ML and REML estimates for these data are given in Table 6.6.

There are three sets of estimates presented in Table 6.6. First the 50 subjects for which there are duplicate measurements are analysed alone. Then the other 43 subjects are included in the analysis. Here, however, there appears to be a problem. The standard errors of the variance estimates are no better after

Table 6.5 Test–retest data for planimetry-based estimates of log(VBR) with missing data

PLAN1	PLAN3	PLAN1	PLAN3	PLAN1	PLAN3
2.05	2.13	1.55	1.53	1.63	1.34
1.72	1.28	2.12	2.30	2.32	*
1.93	1.79	1.69	*	−0.22	*
2.16	1.96	2.03	*	2.09	*
2.27	1.95	1.99	*	1.22	*
2.53	2.17	2.48	*	2.16	*
1.79	1.67	0.99	*	1.84	*
1.87	1.48	2.29	*	1.87	*
1.57	1.57	1.95	*	2.03	*
1.39	1.39	1.89	*	2.27	*
1.89	1.84	1.19	*	1.72	*
2.39	2.26	2.15	*	1.46	0.96
1.67	1.72	1.87	1.41	1.79	1.84
1.57	1.39	2.33	1.84	1.39	1.16
2.30	2.25	1.96	1.53	2.40	2.30
2.03	1.93	2.09	1.89	1.39	1.41
1.19	1.70	2.22	1.89	1.67	1.34
1.13	0.41	2.03	1.46	2.26	2.12
1.63	1.22	1.69	1.63	2.30	2.50
1.93	2.03	2.01	1.50	1.57	1.59
1.89	1.50	2.08	1.55	1.16	*
1.63	2.03	2.13	2.09	1.19	*
1.70	1.96	1.69	1.13	2.00	*
2.82	2.84	2.04	*	1.76	*
0.53	0.99	1.77	*	1.59	*
2.24	2.03	1.79	*	1.53	*
1.63	1.76	2.12	*	1.72	*
2.32	*	1.95	*	1.81	*
1.19	*	0.88	*	1.95	*
1.96	*	1.13	*	2.24	*
1.92	*	1.86	*	0.53	*

* Missing value.

Table 6.6 Maximum likelihood estimates of the variance components for the CAT scan data in Table 6.5

(a) 50 subjects with replicate measurements

Variance	ML estimate (s.e.)	REML estimate (s.e.)
σ_τ^2	0.130 (0.031)	0.133 (0.032)
σ_e^2	0.049 (0.010)	0.049 (0.010)

(b) All 93 subjects

Variance	ML estimate (s.e.)	REML estimate (s.e.)
σ_τ^2	0.169 (0.032)	0.171 (0.032)
σ_e^2	0.050 (0.010)	0.050 (0.010)

(c) Outlier excluded (92 subjects)

Variance	ML estimate (s.e.)	REML estimate (s.e.)
σ_τ^2	0.134 (0.026)	0.136 (0.026)
σ_e^2	0.049 (0.010)	0.049 (0.010)

analysis of the full 93 subjects when compared with the analysis of the 50 without missing data! This appears to be due to a presence of an outlier (-0.22). The third analysis is based on data from 92 subjects (that is, after excluding the outlier). This result is much more satisfactory, but has not led to a particularly marked improvement in the standard error of $\hat{\sigma}_\tau^2$. That of $\hat{\sigma}_e^2$ is, of course, unchanged. The extra 42 subjects contribute no information concerning measurement errors.

For data obtained using a (k_1, k_2)-design Thompson and Anderson (1975) define the following sums of squares:

$$S_1 = \left(\sum_{i=1}^{k_1} X_{i1}^2 \right) - G_1^2/k_1 \tag{6.50}$$

$$S_2 = \left(\sum_{i=k_1+1}^{k} (X_{i1} + X_{i2})^2/2 \right) - G_2^2/2k_2 \tag{6.51}$$

$$S_3 = \sum_{i=k_1+1}^{k_1} (X_{i1} - X_{i2})^2/2 \tag{6.52}$$

$$S_4 = (2k_2 G_1 - k_1 G_2)^2/(2Nk_1 k_2) \tag{6.53}$$

In these expressions

$$G_1 = \sum_{i=1}^{k_1} X_{i1} \quad \text{and} \quad G_2 = \sum_{i=k_1+1}^{k} (X_{i1} + X_{i2})$$

where $k = k_1 + k_2$. The total number of observations is N, where $N = k_1 + 2k_2$. Provided that k_1 is not zero, the expected values of these sums of squares are given by

$$E(S_1) = (k_1 - 1)(\sigma_\tau^2 + \sigma_e^2) \tag{6.54}$$

$$E(S_2) = (k_2 - 1)(2\sigma_\tau^2 + \sigma_e^2) \tag{6.55}$$

$$E(S_3) = k_2 \sigma_e^2 \tag{6.56}$$

$$E(S_4) = \sigma_e^2 + 2k\sigma_\tau^2/N \tag{6.57}$$

Equating sums of squares with their expected values provides the following estimators:

$$\hat{\sigma}_\tau^2 = \{(S_1 + S_2 + S_4)/(k - 1) - S_3/k_2\}/\lambda \tag{6.58}$$

where $\lambda = (N^2 - N - 2k_2)/N(k - 1)$, and

$$\hat{\sigma}_e^2 = S_3/k_2 \tag{6.59}$$

Details of the computation for the CAT scan data in Table 6.5 are presented in Table 6.7.

Table 6.7 Modified analysis of variance for the CAT scan data in Table 6.5

$$G_1 = 74.58 \qquad G_2 = 178.64$$

$$\sum_{i=1}^{k_1} X_{i1}^2 = 140.4408$$

$$\sum_{i=k_1+1}^{k} (X_{i1} + X_{i2})^2/2 = 334.5604$$

$$\sum_{i=k_1+1}^{k} (X_{i1} - X_{i2})^2/2 = 2.453$$

$$s_1 = 140.448 - 74.58^2/42 = 8.0152$$
$$s_2 = 334.5604 - 178.64^2/100 = 15.4379$$
$$s_3 = 2.453$$
$$s_4 = (2(50)(74.58) - 42(178.64))^2/(2(142)(42)(50))$$
$$\quad = 0.0034$$
$$\hat{\sigma}_e^2 = 0.049 \qquad \hat{\sigma}_\tau^2 = 0.135$$

6.6 Two-way mixed effects models

Suppose that, instead of using the methods described in Chapter 5, the data on string lengths given in Table 1.2 is analysed using analysis of variance techniques. It is quite straightforward to carry out a two-way analysis of variance on these data and the results are given in Table 6.8. From the previous analyses of these data it is known that the three raters (Graham, Brian and David) can be regarded as congeneric tests. This implies that the difference between any two ratings of the length of a piece of string will depend on the identity of that

Table 6.8 Analysis of variance for string lengths (Table 1.2)

Source of variation	d.f.	S.S.	M.S.
Rater	2	1.134	0.567
String	14	109.010	7.786
Residual	28	3.146	0.112

Test of equality of rater effects:

$$F_{2,28} = 0.567/0.112$$
$$\quad = 5.063 \quad (p < 0.05)$$

piece of string (this is so because of the inequality of the β_k of Equation (5.1)). This will be expressed in the terminology of an analysis of variance as an *interaction* between rater and string. Unfortunately, since each of the three raters rates each string only once, this interaction will be confounded with the measurement errors. Hence the use of the term 'residual' instead of 'error' in the bottom row of Table 6.8.

The measurement model for the two-way analysis of variance has the form

$$X_{ik} = \mu + \tau_i + \alpha_k + \gamma_{ik} + e_{ik} \tag{6.60}$$

Here μ is an overall mean and the other terms measure departures from that overall mean. The 'effect' of the jth length of string is given by τ_i and that of the kth rater is measured by α_k. The interaction term and measurement error are represented by γ_{ik} and e_{ik}, respectively.

The string effect is a random variable with mean zero and variance σ_τ^2. Either the rater effect can be considered to be a random variable too or α_1, α_2 and α_3 can be considered to be fixed parameters characteristic of Graham, Brian and David, respectively. If the three raters are considered to be a (small) random sample of raters from a larger population of interest, and there is an interest in making inferences concerning this population of raters, then it would be appropriate to consider α_k as a random variable. In this case Equation (6.60) would describe a two-way *random effects* model. In the model given in Section 5.1, however, the three raters were regarded as the only three raters of interest – the α_k being three parameters to be estimated from the data. If the same decision is made here, then Equation (6.60) describes a two-way *mixed effects* model (or simply, a two-way mixed model). It is a mixed model since it contains a mixture of *fixed* effects (μ and α_k) and random effects (τ_i, γ_{ik} and e_{ik}). In both the random effects model and the mixed model the interaction, γ_{ik}, is considered to be a random variable with mean zero and variance σ_γ^2. Similarly the measurement error, e_{ik}, has an expectation of zero and variance σ_e^2. Finally, all random effects are assumed to be uncorrelated.

Considering the mixed model only, the variance of the observations is given by

$$\mathrm{Var}(X_{ik}) = \sigma_\tau^2 + \sigma_\gamma^2 + \sigma_e^2 \tag{6.61}$$

If there are no constraints imposed on the interaction the reliability of any one rater will be given by

$$R = \frac{\sigma_\tau^2}{\sigma_\tau^2 + \sigma_\gamma^2 + \sigma_e^2} \tag{6.62}$$

This is the expression for the reliability coefficient as derived by Bartko (1966). Shrout and Fleiss (1979) consider a two-way mixed model which leads to a derivation of a reliability coefficient that is superficially very different to that in Equation (6.62). This arises from the introduction of the constraint

$$\sum_k \gamma_{ik} = 0 \qquad \text{for all } i \tag{6.63}$$

The results are not contradictory but do act as a warning for the user of computer software to check what arbitrary constraints have been used in the parameterisation of the fitted variance components models. The equivalence of the two ways of estimating variance components are described in detail on pages 400–4 of Searle (1971). In both approaches the mean of the fixed effects (the α_k) is constrained to be zero.

Apart from estimating variance components one also wishes to estimate the fixed effects in a mixed model. One might be interested in testing various hypotheses concerning these fixed effects. Table 6.8, for example, includes a simple F-test of the equality of the three rater effects. If one were interested in exploring differences between raters having already obtained a significant F-test for the rater effects, then it would be advisable to use one of the simultaneous test procedures available (see Fleiss, 1986, or Hand and Taylor, 1987, and also Section 7.2).

Fleiss (1986) uses the two-way mixed model for the analysis of results from randomised blocks and balanced incomplete blocks designs (see Section 4.6). The appropriate measurement model here is

$$X_{ik} = \mu + \tau_i + \alpha_k + e_{ik} \tag{6.64}$$

with $\sum_k \alpha_k = 0$. This is similar to Equation (6.60) except that it is assumed *a priori* that there are no rater–subject interactions (i.e. $\gamma_{ik} = 0$ for all i and k). If one wished to test this assumption then replicate ratings of each subject by each of the raters would be required. In the latter case one would have a situation analogous to a typical laboratory precision study. The example described in Section 4.7 involved a random sample of four laboratories each being sent five samples at each of four concentrations of aqueous solutions of ammonia. The data generated by this experiment can be considered to conform to that of a two-way mixed model. Here laboratories are random and the four ammonia concentrations fixed.

Modern methods of variance components estimation for mixed models will be discussed in the following section. Traditionally, workers have used derivations of expected mean squares and equated them with the corresponding entries for the mean squares in an ANOVA table. This is a reasonably straightforward approach for balanced data (see, for example, Bartko, 1966; Shrout and Fleiss, 1979; or Cronbach *et al.*, 1972) but is a bit tedious for balanced incomplete blocks designs (see Fleiss, 1981a and 1987). The method loses most of its attractions for unbalanced data, however, and for this reason will be rarely referred to in the rest of this chapter. The enthusiast is referred to Searle (1971) or Cronbach *et al.* (1972). Expected mean squares (as linear functions of the required variance components) can be produced as part of the output generated using the commonly available software packages (see Appendix 4) and one useful role that they might play is as a means of checking the form of constraints imposed on the random interaction effects in mixed models (see above).

6.7 The general mixed model

Although the model described by Equation (6.22) was introduced as a one-way random effects model it can still be described as a simple example of a mixed model. The mean (μ) is a fixed effect, the others (τ_i and e_{ij}) are random effects. The models described by Equations (6.60) and (6.64) are also mixed models even when the rater effects are regarded as random. Again the mean (μ) is a fixed effect common to both of these models. In general a linear model might have several fixed effects together with several random effects. The decision concerning whether particular effects are fixed or random might not always be straightforward, but once this has been made, the mathematical form of the model should be clear. The actual form of the model will, of course, depend on the structure displayed by the data. It will take into account both crossing and nesting, as appropriate. It is also possible to introduce the effects of covariates into these general models.

Suppose, for example, that one asks each of a random sample of chemistry students to construct a linear calibration curve of the type described in Section 1.8. Assuming a common slope for all of the students, but allowing for a fixed bias characteristic of each student a suitable model would be

$$X_{ik} = \mu + \alpha_k + \beta\eta_i + e_{ik} \tag{6.65}$$

Here X_{ik} is the measurement produced by student k on the ith concentration of material with *known* value of η_i. In this model β is a fixed parameter, and so is μ. The random effects are represented by α_k and e_{ik}. If the random effects both have expectations of zero and variances given by σ_α^2 and σ_k^2, respectively, then it follows that

$$\underset{k}{E}(X_{ik}) = \mu + \beta\eta_i \tag{6.66}$$

and

$$\text{Var}(X_{ik}) = \sigma_\alpha^2 + \sigma_k^2 \tag{6.67}$$

Furthermore, if random effects are uncorrelated, then

$$\text{Cov}(X_{ik}, X_{jk}) = \sigma_\alpha^2 \quad (i \neq j) \tag{6.68}$$

Measurements made by the same student will be correlated. Note that, although it is usual to specify that σ_k^2 is the same for all k in a variance components model, this has not been done here. The only reason for doing this (other than the fact that it might be true!) is computational convenience.

Now suppose that, in addition to having characteristic fixed biases and different measurement precisions, the students also produce evidence of characteristic slopes rather than a single common one, β. Let the slope for student k be

$$\beta_k = \beta + \gamma_k \tag{6.69}$$

The model described by (6.69) now becomes

$$X_{ik} = \mu + \alpha_k + \beta_k\eta_i + e_{ik} \tag{6.70}$$

$$X_{ik} = \mu + \alpha_k + \beta\eta_i + \gamma_k\eta_i + e_{ik} \tag{6.71}$$

Here the random effects are now α_k, $\gamma_k\eta_i$ and e_{ik}. Making the usual assumptions concerning lack of correlation, the random effects might now need to be modified. It would be reasonable to expect that the slope and intercept of a student's calibration would be correlated. Let $\text{Var}(\gamma_k) = \sigma_\beta^2$ and $\text{Cov}(\alpha_k, \gamma_k) = \sigma_{\alpha\beta}$. Now

$$\underset{k}{E}(X_{ik}) = \mu + \beta\eta_i \tag{6.72}$$

$$\text{Var}(X_{ik}) = \sigma_\alpha^2 + \eta_i^2\sigma_\beta^2 + \sigma_k^2 + 2\eta_i\sigma_{\alpha\beta} \tag{6.73}$$

and

$$\text{Cov}(X_{ik}, X_{jk}) = \sigma_\alpha^2 + \sigma_\beta^2 \eta_i \eta_j + \sigma_{\alpha\beta}(\eta_i + \eta_j) \qquad (i \neq j) \qquad (6.74)$$

Goldstein (1987) describes an equation analogous to (6.71) as an example of a *general multilevel linear model*. In the context of the present chapter, it can be regarded as a slightly complicated example of the general mixed model and, on the whole, the discussion will concentrate on simpler examples. It should be stressed, however, that there is no great theoretical gap between the simple and complex models, but there may be a few computational hurdles to cross in getting from one to the other.

Searle (1987) provides an up-to-date survey of modern estimation methods for the mixed model. These include the traditional ANOVA methods (equating mean squares with their expected values). Henderson's adaptations of the ANOVA methods (Henderson, 1953), *minimum norm quadratic unbiased estimation* (MINQUE; see LaMotte, 1973 or Rao, 1972), maximum likelihood (ML) and restricted maximum likelihood (REML). Two further estimation methods are *minimum variance quadratic unbiased estimation* (MIVQUE; see Hartley *et al.*, 1978) and *iterative generalised least squares* (IGLS; see Goldstein, 1986 and 1987). Harville (1977) provides a detailed review of the properties, advantages and disadvantages of the two likelihood-based methods (ML and REML). Swallow and Monahan (1984) summarise Monte Carlo investigations to compare the properties of ANOVA, MIVQUE, REML and ML estimators of variance components.

ML and REML estimators of variance components were introduced in Section 6.4 and will not be discussed in detail here. The others will not be discussed at all. Briefly, ML estimates the variance components by those values which maximise the full likelihood function over the parameter space. REML, on the other hand, involves the partitioning of the likelihood into two parts, one of which is free of the fixed effects. REML maximises only the part of the likelihood that is free of the fixed effects. Methods of estimation of the fixed effects, given the variance components estimates, are discussed in Searle (1987). It does not appear to be possible at present to say which of ML or REML is the 'best' method of estimation, although REML estimators have less bias. Both methods require extensive computation, particularly in the case of complex designs and/or significant lack of balance. Details of computer software are provided in Appendix 4. Readers interested in Bayesian approaches to estimation are referred to Lee (1987).

6.8 Analysis of nested data

Nested or hierarchical designs for reliability studies were introduced in Section 4.7. This section will simply illustrate some of the properties of the estimators of variance components obtained from nested designs. As above, only ML or REML estimators will be used for the examples. Snedecor and Cochran (1967; pages 286–8) describe the use of ANOVA estimators to analyse the results of an experiment to investigate the variability in measurements of calcium in turnip leaves. Four plants were selected at random and then three leaves were randomly selected from each plant. Finally, two samples of 100 mg were removed from each of the leaves and calcium determined in these samples using microchemical methods. These data can be used to model variation *between* plants, variation between leaves *within* plants, and variation between samples *within* leaves. It should be clear from the sampling design that the estimate of variability for plants will not be very precise – it will be based on only four plants. That for leaf-samples, however, will be much more precise, being based on twelve pairs of measurements.

D. L. Robinson (1987) illustrates the use of REML estimators to analyse variability in measurements of the ratio of two isotopes of nitrogen (^{14}N to ^{15}N) in soil samples. The aim of the experiment was to compare the precision of two different mass spectrometers used to measure the ^{14}N to ^{15}N ratios. Three replicate plots of land previously treated with ^{15}N were chosen and six soil samples taken from each plot. Samples were homogenised and then four subsamples taken from each sample. Two of the four subsamples were analysed using one of the two mass spectrometers (Machine A) and the other two were analysed using the second machine (Machine B). Unlike most traditional analyses of variance components, Robinson's REML analysis allows for heterogeneity in the error variances. The REML estimates of the four variance components were as follows: plots, 63.2 (standard error 99.9); samples, 157.0 (s.e. 60.2); Machine A, 27.2 (s.e. 9.3) and Machine B, 1.7 (s.e. 0.6). The variance components for plots and samples are very clearly estimated with much lower precision than are the measurement error variances for the two mass spectrometers. This does not matter, of course, since the aim of the study was to compare the performance of the two machines and not to look at the variability of soil samples. Note that there is no variance component corresponding to differences (relative bias) between the two spectrometers. Machine effects would have been regarded as a fixed component in the mixed model for these data.

For a third example, reconsider the data and spike and wave activity given in Table 5.17. The durations of 337 episodes of epileptic activity (event) are each measured four times (two raters crossed with two methods of recording). This is an example of a fully crossed design and its analysis will be left as an exercise for the reader. Here events are obviously random, the methods of recording are fixed, and the raters (EEG technicians) could be considered to be either random or fixed (depending on the aims of the investigator). Any sensible analysis of these data will clearly have to allow for heterogeneity in the variances of the errors of measurement (see Exercise 6.1). Here the analysis will be somewhat simplified by consideration of the EEG measurements alone and by regarding the two raters as providing simple replicate measures. Complications will be introduced, however, by the analysis of the data from several children at once. This is illustrated by the data in Table 6.9.

Table 6.9 shows data on the duration of ten events recorded from each of a sample of four children. The duration of each event is measured twice from an EEG chart recording. This is a further example of a simple nested design. Events are nested within children and replicates are in turn nested within events. ML and REML estimates of the three variance components, together with their standard errors, are shown in Table 6.10. These data illustrate the same points as the analyses above except for the added complication of a zero estimate for σ^2_{child}. The latter is also a common characteristic of estimates of variance components arising from nested designs (Leone and Nelson, 1966).

Table 6.9 Replicate measurements of the duration* of episodes of spike and wave activity taken from EEG recordings from four children

Child A Event	Duration	Child B Event	Duration	Child C Event	Duration	Child D Event	Duration
1	25,25	1	40,30	1	50,55	1	40,40
2	105,105	2	25,20	2	40,40	2	40,40
3	0,10	3	135,130	3	30,25	3	40,35
4	160,155	4	40,40	4	20,20	4	40,60
5	40,40	5	10,15	5	20,25	5	60,70
6	15,15	6	60,55	6	40,40	6	40,65
7	90,100	7	25,20	7	20,20	7	20,25
8	20,30	8	15,0	8	70,70	8	15,25
9	50,55	9	10,0	9	20,20	9	40,65
10	80,80	10	30,10	10	100,95	10	60,65

* In units of $\frac{1}{10}$ s.

Table 6.11 gives the results of an analysis of duplicate measurements of episode durations based on ten events from each of ten children (raw data not shown). The changes in precision of the estimates of the three variance components as one passes from the top of the hierarchy to the bottom are very clearly seen. Again, this does not matter if the main aim of the study is to estimate the precision of the method of assessment of the duration of each episode of epileptic activity. Different conclusions might hold, however, if estimation of reliability coefficients is considered. Before moving on to reliability estimates, however, it might be sensible to ask whether the analysis is appropriate. In fitting a nested model to these data one is estimating within-child variability in event durations on the implicit assumption that it is the same for each child. This may not be justified. A similar assumption is also made concerning the variability of measurement errors. It will be left as an exercise for the reader to analyse the data for each child in Table 6.9 separately using a one-way random effects model (see Exercise 6.2). The aim in an analysis of any set of data, whether or not it involves testing, is to describe the data as effectively as possible. This might involve fitting several alternative models (including those described in Chapter 5, if appropriate) and it might also require analyses of different aspects of the data separately. The aim is to understand the structure of the data. It should not simply be an exercise in semi-automated computer modelling and parameter estimation, however sophisticated the required software might be.

Finally, consider the use of reliability coefficients for nested data. Suppose data on school children are being collected with the intention of using the information for comparison of different schools (see, for example, Brennan,

Table 6.10 Estimates of Variance components for EEG data in Table 6.9

Variance component	ML estimate (s.e.)	REML estimate (s.e.)
σ^2_{child}	0	0
σ^2_{event}	1110.6 (253.0)	1139.6 (262.8)
σ^2_{error}	41.6 (9.3)	41.6 (9.3)

Table 6.11 REML estimates of variance components based on EEG data from ten events for each of ten children*

Variance component	Estimate (s.e.)
σ^2_{child}	1386.4 (718.5)
σ^2_{event}	1345.2 (204.5)
σ^2_{error}	52.5 (7.4)

Covariance matrix for estimates

	σ^2_{child}	σ^2_{event}	σ^2_{error}
σ^2_{child}	516203.983		
σ^2_{event}	−4179.606	41809.835	
σ^2_{error}	0	−27.557	55.115

* Duplicate measures for each event (see Table 6.9).

1975). For simplicity select ten schools (of equal size) and then take a random sample of twenty students from each of these schools. Again, to make the discussion simple, assume that the students are all of the same age. Finally, the reading ability of each of the students is assessed by means of a battery of four parallel tests. Consider the kth test for the jth subject in school i. Then, an appropriate nested model is

$$X_{ijk} = \mu + s_i + c_{ij} + e_{ijk} \tag{6.75}$$

Where X_{ijk} is the required test score, μ is an overall mean, s_i is the effect of school i, c_{ij} is the effect of the jth student *within* school i and e_{ijk} is the appropriate measurement error for test k. Let the variance components for schools, children within schools, and measurements errors be represented by σ^2_s, σ^2_c and σ^2_e, respectively. With the usual assumptions regarding lack of correlation of the random effects it follows that

$$\mathrm{Var}(X_{ijk}) = \sigma^2_s + \sigma^2_c + \sigma^2_e \tag{6.76}$$

However, within a single school (i, say) it is clear that

$$\mathrm{Var}(X_{ijk}) = \sigma^2_c + \sigma^2_e \tag{6.77}$$

Within schools, the reliability of a child's individual test score will be given by

$$R_1 = \frac{\sigma^2_c}{\sigma^2_c + \sigma^2_e} \tag{6.78}$$

For the mean of the four test scores, the reliability will be

$$R_4 = \frac{\sigma^2_c}{\sigma^2_c + \sigma^2_e/4} \tag{6.79}$$

And, finally if the mean of all the twenty students' average scores is used as a measure of the school's performance, then the reliability of this school-based index will then be given by

$$R_5 = \frac{\sigma^2_s}{\sigma^2_s + \sigma^2_{c/20} + \sigma^2_{e/80}} \tag{6.80}$$

Now, if one had estimated the three variance components as in the example of the EEG records and found that $\hat{\sigma}_s^2 = 0$, then there would be a problem! Of course, Equation (6.80) is only a sensible index of reliability if there is indeed variation between schools. But even if σ_s^2 is of comparable size to σ_c^2 one also needs a set of data that will yield estimates of σ_s^2, σ_c^2 and σ_e^2 with reasonable precision. A fully nested design produces a precise estimate of σ_s^2 only if there is a reasonable number of schools included in the study. This should be fairly obvious to the reader, but if not, the results in Table 6.11 should help to make the point.

6.9 Generalisability studies

Since generalisability theory as described by Cronbach and his colleagues (see Cronbach *et al.*, 1963) is so intimately concerned with variance components estimation it would seem appropriate to discuss generalisability studies again at this stage of the present chapter. Initially generalisability theorists worked with reliability or generalisability coefficients analogous to those described in this text (see Cronbach *et al.*, 1963), but later gave greater emphasis to the variance components themselves (see Cronbach *et al.*, 1972 or Shavelson and Webb, 1981). This is more in line with the approaches of physical scientists to their measurement problems. A more traditional approach to the use of reliability coefficients can be found in Maxwell and Pilliner (1968).

Perhaps the major contribution of generalisability theory is its emphasis on the multiple sources of measurement error. Accepting this, one is led to the use of estimated variance components (using the methods of either this chapter or Chapter 5) as a way of assessing the relative magnitude of these different sources of variation. But there is a problem. There is a need for large sample sizes, otherwise many of the estimates of variance components will be practically worthless (see Smith, 1978). Furthermore, if one is to study several facets of observation in a single generalisability study then the required sample size is likely to be well beyond the resources of the investigator. Like any other scientific experiment, the successful generalisability or reliability study is likely to be simple and elegant. It should be designed with a simple purpose in mind and carried out on enough observations to provide the appropriate level of statistical power. The recognition that there are multiple sources of error merely forces the investigator to think a bit more carefully about the way in which measurements are to be made and used. It does not solve any of the investigator's problems.

In their review, Shavelson and Webb (1981) remark: 'With the importance and fallibility of estimated variance components so clearly recognized, we find it astonishing that so little attention has been given to this topic in the literature on G theory.' Similar comments could just as equally be made about the use of factor-analytic models (Chapter 5) or the routine use of κ statistics (Chapter 7). The simple, depressing, conclusion is that sample sizes are almost invariably too low.

In designing a reliability study the investigator should bear two statistical problems in mind. The first concerns the number of observations (statistical power) and the second concerns the best way of allocating a fixed number of observations to the different facets of observation (optimal design). In the context of a one-way random effects model, for example, work such as that of Donner and Eliasziw (1987) provides guidelines for sample size requirements. This work, however, does not indicate whether it is better to work with as many subjects as possible (with the necessity of only making very few observations on each subject) or with as many replicate measurements on each of a very few subjects. Should one always aim for a balanced design? Or are there situations where an unbalanced design is better? The reader is referred to Scheffé (1959), Thompson and Anderson (1975) and P. L. Smith (1978) for further discussion of these problems.

6.10 Exercises

6.1   Use variance components models to analyse the EEG data in Table 5.17. Bear in mind that the EEG-based measurements have a lower precision than the Monolog recordings. If the analysis appears to be difficult, try analysing different parts of the data separately or introduce simplifying assumptions (consider, for example, the way in which these data were analysed in Chapter 5).

6.2   Using a one-way variance components model, investigate the sources of variation in EEG measurements in Table 6.9 separately for each of the four children. Comment on the applicability of the use of a nested variance components model for these data.

6.3   Use the delta technique to derive the approximate variance of the reliability coefficient defined by Equation (6.80). Use this result to estimate the standard error of the reliability of a single child's total score (duration of epilepsy) based on the results presented in Table 6.11.

6.4   Consider a randomised blocks experiment involving r raters assessing each of n children (total number of observations, $N = nr$). Derive expected mean squares for the two-way random effects model appropriate for this design and for the corresponding two-way mixed model (raters fixed).

6.5   Compare the uses of covariance-components models (Chapter 5) with those of the variance-components models described in the present chapter. What are the relative advantages and disadvantages of the two approaches?

6.6   Consider the (k_1, k_2) – design described in Section 6.5. Carry out a small simulation study to investigate the effects of varying k_1 and k_2 (fixing $k_1 + 2k_2 = 100$, say) on the sampling variability of (a) the error variance, (b) the variance of the subjects being measured, and (c) the reliability of the measurements. (*Hint:* Start with a true reliability of 0.9.)

6.7   Discuss the following statement by Shavelson and Webb (1981, page 138).

While we consider the problems associated with estimating variance components to be the Achilles heel of G theory, these problems afflict all sampling theories. One virtue of G theory is that it brings estimation problems to the fore and puts them up for examination.

6.8 Write an essay on the title of the 1963 paper by Cronbach *et al.*: 'Theory of Generalizability: A Liberalization of Reliability Theory.'

6.9 Discuss the use of bootstrap and jackknife techniques in estimating standard errors for variance components in two-way random effects models. (See Cronbach *et al.*, 1972, pages 54–57.)

7

Statistical Methods for Categorical Measurements

7.1 Assessment of diagnostic tests

The presence of illness in an individual often cannot be determined with certainty. The use of diagnostic tests of high sensitivity and specificity is clearly very desirable (see Section 1.8) but frequently their routine use is impracticable because they are too expensive. In addition, many highly accurate diagnostic tests might be invasive and potentially hazardous. Accordingly clinicians and epidemiologists usually prefer to use simpler but more error-prone screening tests. An example from psychiatry is the General Health Questionnaire or GHQ (Goldberg, 1972; see also Section 1.8). The properties of the GHQ are usually assessed by comparison with the results of a detailed semi-structured psychiatric interview (see Tarnopolsky *et al.*, 1979). For the purpose of these validation studies the findings of the psychiatric interview are regarded as being error-free (i.e. the 'truth'). The latter assumption may be unrealistic. In general, if the reference test or 'gold standard' is significantly inaccurate then the estimates of a new screening test's characteristics will be biased. This and other sources of bias in the assessment diagnostic tests are discussed in detail by Begg (1987). Here a more appropriate statistical model will be introduced. This will allow the simultaneous assessment of two or more diagnostic tests in the absence of the required gold standard. This model is an example of a *latent structure* or *latent class model* (see Lazarsfeld and Henry, 1968 or Bartholomew, 1987). It is quite common in epidemiological surveys to use one or more screening tests on samples from two or more populations. A simple example, again from psychiatry, is the use of the GHQ to compare psychiatric morbidity in men and women (see Section 1.9). In general, we are concerned with the results of applying K tests to samples from G populations. Here a model from the results of using two screening tests on samples from two populations will be considered as a relatively simple example.

Let a standard test (Test 1) and a new test (Test 2) be applied simultaneously to each individual in samples from two populations. Let the size of the gth sample ($g = 1, 2$) be N_g and let the probability of being ill in the gth sample by π_g. Let α_{gh} and β_{gh} represent the false positive and false negative rates for test h ($h = 1, 2$) in a sample from population g ($g = 1, 2$), respectively.

The specificity of test h when used on samples from population g is $1 - \alpha_{gh}$ and, similarly, its sensitivity is $1 - \beta_{gh}$. Finally, it is assumed that conditional on the true state of an individual (ill or well), the two tests are statistically independent. That is, they have independent errors. This *assumption of conditional independence* is a crucial component of this and the other measurement models presented in this text. It is important that the investigator is convinced of the truth of this assumption before proceeding to the use of a latent class model as described below. Diagnostic tests based on totally different procedures (use of X-rays versus use of blood tests, for example) may well be conditionally independent but two psychiatric screening questionnaires containing overlapping items will not be.

Under the above assumptions, the frequencies of the possible test outcomes in the two populations are described by two multinomial distributions (see Exercise A1.5). Let P_{gij} be the observed proportion of sample g with test outcomes i and j in Tests 1 and 2, respectively. Let $i = 1$ if Test 1 is positive and $i = 2$ if it is negative. Similarly, let $j = 1$ if Test 2 is positive and $j = 2$ if Test 2 is negative. The four possible pairs of test outcomes for group g, together with their probabilities of occurrence, are given in Table 7.1.

Table 7.1 Results of applying two diagnostic tests to group g *

Outcome		Joint probability
Test 1	Test 2	
1	1	$\pi_g(1 - \beta_{g1})(1 - \beta_{g2}) + (1 - \pi_g)\alpha_{g1}\alpha_{g2}$
1	2	$\pi_g(1 - \beta_{g1})\beta_{g2} + (1 - \pi_g)\alpha_{g1}(1 - \alpha_{g2})$
2	1	$\pi_g\beta_{g1}(1 - \beta_{g2}) + (1 - \pi_g)(1 - \alpha_{g1})\alpha_{g2}$
2	2	$\pi_g\beta_{g1}\beta_{g2} + (1 - \pi_g)(1 - \alpha_{g1})(1 - \alpha_{g2})$

*See text for definitions of the parameters.

The joint likelihood for the observations from the two samples, ignoring constant terms, is given by

$$l = \prod_{g=1}^{2} \{\pi_g(1 - \beta_{g1})(1 - \beta_{g2}) + (1 - \pi_g)\alpha_{g1}\alpha_{g2}\}^{N_g P_{g11}}$$
$$\times \{\pi_g(1 - \beta_{g1})\beta_{g2} + (1 - \pi_g)\alpha_{g1}(1 - \alpha_{g2})\}^{N_g P_{g12}}$$
$$\times \{\pi_g\beta_{g1}(1 - \beta_{g2}) + (1 - \pi_g)(1 - \alpha_{g1})\alpha_{g2}\}^{N_g P_{g21}}$$
$$\times \{\pi_g\beta_{g1}\beta_{g2} + (1 - \pi_g)(1 - \alpha_{g1})(1 - \alpha_{g2})\}^{N_g P_{g22}} \qquad (7.1)$$

Parameter estimates can now be obtained by maximising this likelihood function, but there is a problem. In general, for K tests applied to G populations, there are $(2K - 1)G$ degrees of freedom for estimating $(2K + 1)G$ parameters. In the present example there are 6 degrees of freedom and ten parameters ($\pi_1, \pi_2, \beta_{11}, \beta_{12}, \beta_{21}, \beta_{22}, \alpha_{11}, \alpha_{12}, \alpha_{21}, \alpha_{22}$). This model is clearly under-identified (over-parameterised). The general question of identifiability of latent class models is discussed by McHugh (1956) and Goodman (1974) but clearly a necessary condition for the current model to be identified is that

$$(2^K - 1)G \geq (2K + 1)G \tag{7.2}$$

or

$$2^K \geq 2(K + 1) \tag{7.3}$$

$$K \geq 3 \tag{7.4}$$

The case for using three raters (or tests) in the measurement of the reliability of clinical data has been made by Walter (1984). Hui and Walter (1980) introduce constraints on the parameters so that it is not over-parameterised when using only two diagnostic tests. They introduce the constraints

$$\left. \begin{array}{l} \alpha_{gh} = \alpha_h \\ \beta_{gh} = \beta_h \end{array} \right\} \quad \text{for all } g \tag{7.5}$$

The number of parameters to be estimated is now given by $2R + G$ and the number of degrees of freedom is greater than the number of parameters if

$$G \geq \frac{K}{(2^{K-1} - 1)} \tag{7.6}$$

For $K = 2$ then one can apply the model to data for two groups, and when $K = 1$ then, as before, one can apply the data to a sample for a single population.

Numerical methods of maximising the likelihood for the latent class model for two tests used on two groups are discussed by Hui and Walter (1980). The above model is still not identified and these authors also introduce the constraint

$$1 - \alpha_1 - \beta_1 > 0$$

so that it is. This is a reasonable assumption for the standard test if it is any practical value. General methods of fitting data to latent class models, and, in particular, the use of the EM algorithm, are described by Everitt and Hand (1981), Everitt (1984) and Bartholomew (1987).

Hui and Walter (1980) illustrate their latent class model and estimation method using the data given in Table 7.2. In this example the new Tine test (Test 2) was evaluated against the standard Mantoux test (Test 1) for the detection of tuberculosis. Data for Population 1 were derived from Greenberg and Jekel's (1969) study in a southern US school district and data for Population 2 were from that of the Missouri State Sanatorium (Capobres *et al.*, 1962). The maximum likelihood estimates and their standard errors (in brackets) were found to be:

$$\hat{\alpha}_1 = 0.0067 \ (0.0038)$$

$$\hat{\alpha}_2 = 0.0159 \ (0.0056)$$

$$\hat{\beta}_1 = 0.0339 \ (0.0069)$$

$$\hat{\beta}_2 = 0.0312 \ (0.0062)$$

$$\hat{\pi}_1 = 0.0268 \ (0.0071)$$

$$\hat{\pi}_2 = 0.7168 \ (0.0128)$$

Table 7.2 Results of Mantoux and Tine tests for tuberculosis in samples from two populations

| | Population 1 | | |
| | Tine test | | |
Mantoux test	Positive	Negative	Total
Positive	14	4	18
Negative	9	528	537
Total	23	532	555
	Population 2		
	Tine test		
Mantoux test	Positive	Negative	Total
Positive	887	31	918
Negative	37	367	404
Total	924	398	1322

Reproduced with permission from Hui and Walter (1980).

Hui and Walter were able to partly check the validity of their latent class model for Population 2 by comparing their estimates of β_1 and β_2 to those of Capobres *et al.* (1962), who applied the two diagnostic tests to patients with pulmonary tuberculosis (as verified by a positive culture of acid-fast bacilli).

Now consider a second example. The data arise from histological assessments of 118 biopsy slides by seven independent pathologists. The data are adapted from the study of Holmquist *et al.* (1967) which has been re-analysed by Landis and Koch (1977b) and others. The study was designed to investigate the variability in the histological classification of carcinoma *in situ* and other related lesions of the uterine cervix. The original coding used by the seven pathologists was: (1) negative, (2) atypical squamous hyperplasia, (3) carcinoma *in situ*, (4) squamous carcinoma with early stromal invasion, and (5) invasive carcinoma. Following Landis and Koch (1977b) the categories were combined in the following way: (0) derived from original categories (1) and (2), and (1) derived from the original categories (3)–(5). The two new codes correspond to absence and presence of carcinoma, respectively. The recoded data are shown in Table 7.3. The proportions of the 118 slides perceived as displaying carcinoma by each of the seven pathologists (A to G, in alphabetical order) are:

$$0.559, \ 0.669, \ 0.381, \ 0.271, \ 0.602, \ 0.212, \ 0.559$$

Pathologists A, B, E and G appear to have reasonably similar marginal proportions (average 0.597). Pathologists C, D and F also show some similarities (average 0.288), but it is not at all clear from these marginal proportions which group appear to be 'better' at detecting signs of carcinoma.

One can now fit a two-latent class model using the EM algorithm (utilising a program written by Professor B. S. Everitt) to produce maximum likelihood estimates for the proportion of slides displaying evidence of carcinoma (π) and the seven pairs of false positive and false negative rates (one pair for each

Table 7.3 Independent histological assessments by seven pathologists of the presence (1) or absence (0) of cancer of the cervix*

Rater							Rater						
A	B	C	D	E	F	G	A	B	C	D	E	F	G
1	1	1	0	1	1	1	0	1	0	0	1	0	1
0	0	0	0	0	0	0	1	1	1	1	1	1	1
1	1	1	1	1	1	1	1	1	1	1	1	0	1
1	1	1	1	1	1	1	0	0	0	0	0	0	0
1	1	1	1	1	1	1	0	1	0	0	1	0	0
0	0	0	0	0	0	0	1	1	0	1	1	0	1
0	0	0	0	0	0	0	0	0	0	0	0	0	0
1	1	0	1	0	0	1	1	1	1	1	1	1	1
0	0	0	0	1	0	0	1	1	0	1	1	0	1
0	0	0	0	0	0	0	1	1	1	1	1	0	1
1	1	1	1	1	1	1	1	1	0	1	1	0	1
0	0	0	0	0	0	0	0	0	0	0	0	0	0
1	1	1	0	1	1	1	0	0	0	0	0	0	0
0	0	0	0	0	0	0	0	1	0	0	1	0	0
1	1	1	0	1	0	1	0	0	0	0	0	0	0
1	1	0	1	1	1	1	1	1	1	0	1	0	1
0	1	0	0	1	0	1	0	0	0	0	0	0	0
0	0	0	0	0	0	0	1	1	1	1	1	1	1
0	1	0	0	0	0	1	1	1	0	1	1	1	1
0	0	0	0	0	0	0	0	1	0	0	0	0	0
1	1	1	1	1	1	1	1	1	1	0	1	0	1
0	0	0	0	0	0	0	1	1	1	0	1	0	1
0	0	0	0	0	0	0	1	1	0	0	1	0	0
0	0	0	0	0	0	0	1	1	1	0	1	0	1
1	1	1	0	1	1	1	0	0	0	0	0	0	0
1	1	1	0	1	0	1	1	1	1	0	1	1	1
1	1	1	1	1	0	1	1	1	0	0	0	0	0
0	0	0	0	0	0	0	0	1	0	0	1	0	0
1	1	1	1	0	0	1	1	1	1	1	1	0	1
1	1	1	1	1	1	1	0	1	0	0	1	0	0
0	0	0	0	0	0	0	1	1	0	0	1	0	1
1	1	1	0	1	0	1	1	1	0	0	1	0	1
0	0	0	0	1	0	0	0	0	0	0	0	0	0
1	1	0	0	1	0	1	0	0	0	0	0	0	0
1	1	1	1	1	0	1	1	1	0	0	0	0	1
0	0	0	0	0	0	0	1	1	0	0	0	0	0
1	1	0	0	1	0	1	0	1	0	0	0	0	0
1	1	1	1	1	0	1	1	1	0	0	1	0	1
1	1	1	1	1	1	1	0	0	0	0	0	0	0
1	1	1	0	1	0	1	1	1	1	0	1	0	1
1	0	0	0	0	0	0	1	1	0	0	1	0	1
0	0	0	0	0	0	0	0	0	0	0	0	0	0
0	1	0	0	1	0	1	0	0	0	0	0	0	0
1	1	1	1	1	1	1	0	0	0	0	0	0	0
1	1	1	0	1	0	1	1	1	1	0	1	1	1
1	0	0	0	0	0	0	1	1	1	0	1	0	1
0	1	0	0	0	0	0	0	0	0	0	0	0	0
1	1	1	1	1	0	1	0	1	0	0	0	0	0
1	1	1	1	1	1	1	1	1	1	0	1	0	1
1	1	0	0	1	0	1	1	1	0	0	1	0	1
1	1	1	1	1	0	1	0	0	0	0	0	0	0
0	0	0	0	0	0	0	0	0	0	0	0	0	0
0	1	0	0	1	0	1	0	0	0	0	0	0	0
0	0	0	0	0	0	0	1	1	1	0	1	1	1
1	1	1	1	1	1	1	1	1	1	0	1	0	1
0	0	0	0	0	0	0	0	0	0	0	0	0	0
0	1	0	0	0	0	0	0	1	0	0	0	0	0
1	1	1	1	1	0	1	0	0	0	0	0	0	0
0	1	0	0	0	0	0	0	1	0	0	0	0	0

* Adapted from Holmquist et al. (1967).

pathologist). The results are shown in Table 7.4. The first group of pathologists (A, B, E and G) appear to be detecting some carcinoma when it is not present but also missing very few true carcinomas. The second group (C, D and F) are producing few false positives but are in turn missing quite a high proportion of the slides with carcinoma. The reader will be left to decide which pattern is 'better'.

Table 7.4 Maximum likelihood estimates for a two-class model fitted to the biopsy data presented in Table 7.3

Rater	$\hat{\pi} = 0.501$ $\hat{\alpha}$	$\hat{\beta}$
A	0.116	0.000
B	0.354	0.017
C	0.000	0.239
D	0.000	0.459
E	0.223	0.021
F	0.000	0.577
G	0.116	0.000

Further examples of the use of latent class analysis in medicine can be found in Walter (1984), Rindskopf and Rindskopf (1986), and in Dawid and Skene (1979). Dayton and Macready (1983) describe the use of a simple latent class model to investigate 'error' rates in the classification of children with respect to their ability to conserve weight under shape transformations. Children are presented with a series of tasks designed to assess conservation of weight (Piaget, 1968) and the response to each task is scored 1 for non-conserving or 2 for conserving. The latent variable has two classes corresponding to the 'true' state of the child. In this context Dayton and Macready refer to false negatives and false positives as *'omission'* and *'intrusion'* errors, respectively. Dayton and Macready discuss strategies for fitting a hierarchical series of models to progressively introduce more constraints on the parameter values. They also point out problems arising from the use of small sample sizes. This point is also reinforced by Bartholomew (1987).

7.2 Tests of observer bias

It was seen in the last section that the seven pathologists clearly differed in their probability of assessing a biopsy slide as indicating the presence of carcinoma. The pathologists differed in both their sensitivities and specificities. If it is of interest to test whether these differences are statistically significant then there are two general approaches that might be used. The first involves the introduction of constraints into the appropriate latent class model (equal sensitivities and/or equal specificities amongst some or all of the pathologists, for example) and test the fit of the model using the log-likelihood ratio criterion (see Dayton and Macready, 1983). The second approach involves testing the equality of the observed marginal proportions (for example, the proportions of the 118 perceived as displaying evidence of carcinoma by the seven pathologists). The appropriate test of marginal homogeneity for binary data is the use of Cochran's Q statistic (Fleiss, 1965). This is derived as follows.

Let X_{ik} represent the assessment of the ith slide by the kth pathologist ($i = 1$ to I, $k = 1$ to K) with $X_{ik} = 1$ if the ith slide is judged by the kth observer to be indicative of carcinoma, 0 otherwise. Let X_{i+} represent the total number of observers who judge the ith slide to indicate the presence of carcinoma and, similarly, let X_{+j} represent the total number of slides the jth pathologist judges to have indications of carcinoma. Finally, let X_{++} represent the total number of slides for which carcinoma is judged to be present. Cochran's Q statistic is then given by

$$Q = \frac{K(K-1) \sum_{j=1}^{K} (X_{+j} - X_{++}/K)^2}{KX_{++} - \sum_{i=1}^{I} X_{i+}^2} \tag{7.7}$$

If the null hypothesis of marginal homogeneity is true then, for large samples, Q will be approximately distributed as a chi-square with $K - 1$ degrees of freedom. For the data given in Table 7.3, $Q = 181.59$ with 6 degrees of freedom.

In the case of just two raters (i.e. $K = 2$) Cochran's Q test is equivalent to the well-known McNemar test for the comparison to two correlated proportions (McNemar, 1947).

Using the same terminology as in the general two-way contingency table (see Table 2.9) the observed counts for subjects jointly classified by two raters are shown in Table 7.5. McNemar's test statistic for the comparison of the marginal probabilities (p_1 and p_2) is given by

$$X^2 = \frac{(n_{12} - n_{21})^2}{n_{12} + n_{21}} \tag{7.8}$$

For large samples and on the assumption of marginal homogeneity this statistic will be distributed as a chi-square with one degree of freedom. This is illustrated in Table 7.6 using the two-way table for pathologists A and B from Table 7.3. The difference between p_1 and p_2 for this table is clearly statistically significant (using a significance level, α, of 0.05). However, one needs to recognise that there are 21 possible pairwise comparisons of the seven pathologists in Table 7.3. Each one generates a separate McNemar statistic. These are shown in Table 7.7.

Consider the possibility of Type I errors arising from the use of McNemar's X^2-statistic. Let these errors for the 21 paired comparisons in Table 7.7 be represented by $E_1, E_2, \ldots, E_{21}$. Then the overall probability of making a

Table 7.5 Two-way contingency table for assessment of bias and agreement between two binary measurements

Observed counts		Rater 2 category		
		1	2	Total
Rater 1 category	1	n_{11}	n_{12}	n_{1+} $p_1 = n_{1+}/n$
	2	n_{21}	n_{22}	n_{2+} $q_2 = n_{2+}/n$
	Total	n_{+1}	n_{+2}	n
		$p_1 = n_{+1}/n$	$q_2 = n_{+2}/n$	

Table 7.6 Joint assessment of biopsy slides by Pathologists A and B (see Table 7.3)

		Pathologist B		
		0	1	Total
Pathologist A	0	36	16	52
	1	3	63	66
	Total	39	79	118

McNemar's test: $\quad X^2 = \dfrac{(16-3)^2}{16+3} = 8.895$

Type I error is given by

$$\alpha^* = P(E_1 \text{ or } E_2 \text{ or } E_3 \text{ or } \dots \text{ or } E_{21}) \qquad (7.9)$$

Here the Type I error is to decide that there is at least one of the pair comparisons displaying evidence of relative bias when there are actually no differences between the raters. In a sensible interpretation of the results given in Table 7.7 there is a need to select a significance level (α) for an individual test so that $\alpha^* = 0.05$, for example. To do this use can be made of the Bonferroni inequality (see Fleiss, 1986):

$$P(E_1 \text{ or } E_2 \text{ or } E_3 \text{ or } \dots \text{ or } E_{27}) \le \sum_{i=1}^{21} P(E_i) \qquad (7.10)$$

or

$$\alpha^* \le 21\alpha \qquad (7.11)$$

Replacing the inequality with an equality shows that the appropriate α is approximately 0.0024 (with a corresponding critical value of a chi-square variate with one degree of freedom of about 9.5). Returning to the test statistics given in Table 7.7 it is clear that several of these are significant when using an α of 0.0024.

Table 7.7 McNemar X^2-statistics for the 21 paired comparisons of the seven pathologists in Table 7.3*

		Pathologist						
		B	E	A	G	C	D	F
Pathologist	B	—						
	E	4.57	—					
	A	8.90	1.47	—				
	G	11.27	2.27	0	—			
	C	32.11	26.00	21.00	21.00	—		
	D	47.00	37.10	34.00	34.00	6.76	—	
	F	54.00	46.00	41.00	41.00	14.29	2.56	—

* The pathologists have been re-ordered according to the overall proportion of biopsy slides perceived to be showing evidence of carcinoma (from B, the highest, to F, the lowest).

More general methods of comparing the marginal distributions of categorical assessments made by two or more raters are described by Landis and Koch (1977a). These are based on theoretical work explained in detail in Grizzle *et al.* (1969), or Appendix 1 of Koch *et al.* (1977). Appropriate computer software is described in Appendix 4 of the present book. A number of authors have generalised McNemar's test to the comparison of the marginal frequencies of a $C \times C$ contingency table (see Table 2.9). A test statistic based on the method of moments has been derived by Stuart (1955) and Maxwell (1970). Its properties are discussed in detail by Fleiss and Everitt (1971). As in Table 2.9, let n_{ij} denote the number of observations in the ith row and the jth column and define

$$n_{i+} = \sum_{j=1}^{C} n_{ij} \tag{7.12}$$

$$n_{+j} = \sum_{i=1}^{C} n_{ij} \tag{7.13}$$

$$n = \sum_{i=1}^{C} \sum_{j=1}^{C} n_{ij} \tag{7.14}$$

Further, define

$$d_i = n_{i+} - n_{+i} \tag{7.15}$$

Under the null hypothesis that there are no relative biases between the two raters then it can be shown (see Fleiss and Everitt, 1971) that consistent, but not maximum-likelihood, estimates of the variances and covariances are

$$\hat{\text{Var}}(d_i) = n_{i+} + n_{+i} - 2n_{ii} \tag{7.16}$$

and

$$\hat{\text{Cov}}(d_i, d_j) = -(n_{ij} + n_{ji}) \tag{7.17}$$

Let $\mathbf{d}$ represent the vector of d_i in which one of the d_i has been dropped (it does not matter which one is missed out), and let $\mathbf{V}$ be the covariance matrix for the vector $\mathbf{d}$. The Stuart–Maxwell statistic is then given by

$$X^2 = \mathbf{d}'\mathbf{V}^{-1}\mathbf{d} \tag{7.18}$$

If n is large then X^2 will be distributed as a chi-square distribution with $C - 1$ degrees of freedom. For $C = 3$, Equation (7.18) is equivalent to

$$X^2 = \frac{\bar{n}_{23}d_1^2 + \bar{n}_{13}d_2^2 + \bar{n}_{12}d_3^2}{2(\bar{n}_{12}\bar{n}_{23} + \bar{n}_{12}\bar{n}_{13} + \bar{n}_{13}\bar{n}_{23})} \tag{7.19}$$

where

$$\bar{n}_{ij} = (n_{ij} + n_{ji})/2 \tag{7.20}$$

When C is greater than 3 explicit expressions for X^2 such as Equation (7.19) are probably not worth the computational effort to calculate, although Fleiss and Everitt (1971) provide an explicit expression for the case when $C = 4$. These authors also provide simple numerical examples for the use of the Stuart–Maxwell statistic.

7.3 Variance components models for binary data

This section is concerned with drawing inferences about agreement between dichotomous judgements. It will develop some of the ideas introduced in Sections 2.8 and 2.9 which introduced chance-corrected agreement indices (including Cohen's κ and Scott's π) and intra-class correlations, repectively. In particular, the section will be concerned with the use of analysis of variance (ANOVA) procedures for the estimation of agreement and reliability coefficients.

Table 7.8 shows an analysis of variance table for the dichotomous ratings given in Table 2.1. The latter table provides two examiners' ratings (P and F) for each of 29 candidates. The ANOVA has been carried out just as if the

Table 7.8 Analysis of variance (ANOVA) for dichotomous judgements recorded in Table 2.8

(a) One-way ANOVA

Source of variation	Sum of squares	d.f.	Mean square
Candidates	10.9310	28	0.3904
Residual	3.5000	29	0.1207

(b) Two-way ANOVA

Source of variation	Sum of squares	d.f.	Mean square
Candidates	10.9310	28	0.3904
Examiners	0.8448	1	0.8448
Error	2.6552	28	0.0948
$\hat{\sigma}^2_{Ca} = 0.1478$	$\hat{\sigma}^2_{Ex} = 0.0259$		$\hat{\sigma}^2_{Er} = 0.0948$

results were interval scores rather than binary ratings. Part (a) of Table 7.8 shows the one-way analysis, and Part (b) shows the corresponding two-way analysis. (Variance components have been estimated by simply equating observed and expected mean squares.) If one makes the *a priori* assumption of no examiner bias, then the one-way ANOVA mean squares can provide an estimate of reliability using the expression given in (6.36). That is

$$\hat{\rho}_{1W} = \frac{0.3904 - 0.1207}{0.3904 + 0.1207}$$

$$= 0.528 \tag{7.21}$$

For a reasonably large number of subjects (candidates), expression (6.36) is approximately the same as

$$R_1 = \frac{S - (O + E)}{S + (O + E)} \tag{7.22}$$

where O is the sum of squares due to examiners, S is the sum of squares due to candidates, and, finally, E is the error sum of squares. Substituting for these sums of squares the values given in Table 7.8(b) one obtains

$$R_1 = \frac{10.9310 - (0.8448 + 2.6552)}{10.9310 - (0.8448 + 2.6552)}$$

$$= 0.515 \tag{7.23}$$

Using the terminology for the cell counts and marginal proportions in Table 7.5 it is straightforward to show that R_1 is also given by

$$R_1 = \frac{4(n_{11}n_{22} - n_{12}n_{21}) - (n_{12} - n_{21})^2}{(n_{1+} + n_{+1})(n_{2+} + n_{+2})} \tag{7.24}$$

$$= \frac{4(120) - 49}{(27)(31)}$$

$$= 0.515$$

Equation (7.24) is an alternative expression for Scott's π, the chance-corrected coefficient of agreement described in Section 2.9. The equivalence of R_1 and Scott's π is demonstrated by Fleiss (1975). The assumption of marginal homogeneity in the calculation of Scott's π coefficient is equivalent to that of ignoring examiner biases and using a one-way ANOVA model for the derivation of an intra-class reliability coefficient.

If one were not prepared to assume the lack of examiner bias, then the appropriate model would either be a two-way random effects ANOVA or a two-way mixed model. It will be assumed that there is no candidate–examiner interaction effect. Taking the random effects model first, the appropriate intra-class correlation can be estimated by

$$\hat{\rho}_{2\text{WR}} = \frac{\hat{\sigma}^2_{\text{Ca}}}{\hat{\sigma}^2_{\text{Ca}} + \hat{\sigma}^2_{\text{Ex}} + \hat{\sigma}^2_{\text{Er}}} \tag{7.25}$$

where $\hat{\sigma}^2_{\text{Ca}}$, $\hat{\sigma}^2_{\text{Ex}}$ and $\hat{\sigma}^2_{\text{Er}}$ are estimated variance components for candidates, examiners and error, respectively (see Table 7.8(b)). Here

$$\hat{\rho}_{2\text{WR}} = \frac{0.1478}{0.1478 + 0.0259 + 0.0948} \tag{7.26}$$

$$= 0.551$$

Fleiss (1975) demonstrates that (7.26) is approximately the same as

$$R_2 = \frac{S - E}{S + E + 2O} \tag{7.27}$$

where S, E and O are the sums of squares as defined for Equation (7.22). Here

$$R_2 = \frac{10.9310 - 2.6552}{10.9310 + 2.6552 + 2(0.8448)}$$

$$= 0.542 \tag{7.28}$$

An alternative expression for (7.27) is given by

$$R_2 = \frac{2(n_{11}n_{22} - n_{12}n_{21})}{(n_{1+}n_{+2} + n_{+1}n_{2+})} \tag{7.29}$$

$$= \frac{2(120)}{(323 + 120)}$$

$$= 0.542$$

Equation (7.29) is equivalent to Cohen's κ coefficient.

Finally, considering a mixed effects ANOVA,

$$\hat{\rho}_{2\text{WM}} = \frac{0.1478}{0.1478 + 0.0948} \tag{7.30}$$

$$= 0.609$$

This is the same as

$$R_3 = \frac{S - E}{S + E} \tag{7.31}$$

which in turn can be shown to be identical to

$$R_3 = \frac{2(n_{11}n_{22} - n_{12}n_{21})}{(n_{1+}n_{2+} + n_{+1}n_{+2})} \tag{7.32}$$

The coefficient defined by (7.32) is equivalent to a statistic r_{11} described by Maxwell and Pilliner (1968).

Returning to the assessments of biopsy samples by the seven pathologists in Table 7.3, it is possible to calculate agreement statistics for each of the 21 pairs of pathologists. These agreement statistics can be any of the alternatives prescribed above, although the one that is usually calculated is Cohen's κ (allowing for marginal heterogeneity amongst the pathologists). The calculation of these agreement statistics is presented in Table 7.9.

One might, however, wish to have a single measure of pairwise agreement for the seven pathologists. One obvious candidate for this is the mean of the 21 pairwise agreement coefficients (Light, 1971; Conger, 1980a). For the biopsy data the mean κ value is 0.529. An alternative approach is via analysis of variance. Table 7.10 provides the results of a two-way analysis of variance for the biopsy data in Table 7.3. The estimate of the intra-class correlation for the two-way random effects model is 0.522. Explicit calculation of κ-like statistics measuring overall pairwise agreement between raters in discussed in the following section. Other ANOVA approaches are described in Landis and Koch (1977a, 1977c) and in Fleiss and Cuzick (1979).

Before leaving this section, brief mention must be made of variance components derived from latent trait modelling. These may be more appropriate than those described above in cases where a simple yes/no rating is insufficient to describe how a rater feels about an assessment. Interested readers are referred to the paper by Anderson and Aitkin (1985) for further details. Schwartz (1986) may also be of interest.

Table 7.9 Pairwise agreement statistics for the seven pathologists in Table 7.3

Pathologist pair	P_o	P_e	$P_o - P_e$	$1 - P_e$	κ
AB	0.8390	0.5201	0.3189	0.4799	0.6645
AC	0.8220	0.4859	0.3361	0.5141	0.6538
AD	0.7119	0.4729	0.2390	0.5271	0.4534
AE	0.8559	0.5121	0.3438	0.4879	0.7047
AF	0.6525	0.4658	0.1867	0.5342	0.3497
AG	0.8983	0.5070	0.3913	0.4930	0.7937
BC	0.6949	0.4598	0.2351	0.5402	0.4352
BD	0.6017	0.4224	0.1793	0.5776	0.3104
BE	0.8814	0.5345	0.3469	0.4655	0.7452
BF	0.5424	0.4023	0.1401	0.5977	0.2346
BG	0.8729	0.5201	0.3528	0.4799	0.7352
CD	0.7881	0.5543	0.2338	0.4457	0.5246
CE	0.7797	0.4759	0.3038	0.5241	0.5797
CF	0.7627	0.5684	0.1943	0.4316	0.4502
CG	0.8220	0.4859	0.3361	0.5141	0.6538
DE	0.6525	0.4535	0.1990	0.5465	0.3641
DF	0.8390	0.6319	0.2071	0.3681	0.5626
DG	0.7119	0.4729	0.2390	0.5271	0.4534
EF	0.6102	0.4414	0.1688	0.5586	0.3022
EG	0.9068	0.5121	0.3947	0.4879	0.8090
FG	0.6525	0.4658	0.1867	0.5342	0.3495
		Total	5.5333	10.6350	

Table 7.10 Two-way analysis of variance for biopsy data in Table 7.3*

Source of variation	Sum of squares	d.f.	Mean square
Pathologists	22.10	6	3.6833
Biopsy slides	119.50	117	1.0214
Error	63.94	702	0.0911

$$\hat{\sigma}_p^2 = 0.0304$$
$$\hat{\sigma}_s^2 = 0.1329$$
$$\hat{\sigma}_e^2 = 0.0911$$

* σ_p^2, σ_s^2 and σ_e^2 represent components of variance due to pathologists, biopsy slides and error, respectively.

7.4 Measures of agreement for multinomial data

Consider Table 7.11. This contains data on the assessment of 181 abstracts submitted for consideration for presentation at a statistical conference. Each abstract is assessed by up to eight independent referees and given one of five grades. These grades are A, strongly recommended for presentation as a verbal contribution to the conference; B, recommended as a verbal contribution; C, recommended as a short communication to be presented in the form of a poster; D, recommended for rejection; and E, the referee has no opinion on the suitability of the proposed presentation. Grades A–D can be considered as an ordinal scale; E is here treated as a missing value. It is assumed that missing values occur completely at random.

The overall (marginal) frequencies for each of the grades as used by the eight referees are summarized in Table 7.12. Formal tests for bias will be left as an exercise for the reader (see Exercise 7.1), but it should be fairly obvious without any tests of significance that the referees have differing standards. This is illustrated particularly clearly by the variation in the use of the recommendation for rejection (grade D). Having decided that there is clear evidence of marginal heterogeneity, the discussion will now move on to the assessment of agreement using Cohen's κ statistic and extensions of this statistic for agreement between more than two raters.

Simple κ statistics can be computed for each pair of referees. There are in total 28 of these κ to be calculated and their mean value is 0.104 (range -0.053 to 0.296). By any standards there is very little evidence of good agreement between these referees! Standard errors of the above κ statistics can be calculated using the δ technique (Fleiss *et al.*, 1969) or the jackknife (Schouten, 1986). One could also carry out a similar exercise using one of the alternative weighted κ.

The 28 κ statistics calculated for the data in Table 7.11 involved, for each pair of referees, using only the abstracts that had been assessed by both of the referees. Abstracts with missing data were ignored. Schouten (1985, 1986) has suggested that, in cases where there are missing data, the marginal proportions should be estimated for *all* of the data available for a given rater or referee. Consider assessments by Referee 2 and Referee 3, for example (see Table 7.13). κ can be estimated from this table using the observed agreement (P_o) together with chance agreement (P_e) calculated from the marginal frequencies. That is,

$$P_o = 45/109 = 0.4128$$

$$P_e = \frac{(18)(4) + (38)(32) + (43)(71) + (10)(2)}{109^2}$$

$$= 0.3671$$

$$\kappa = 0.072$$

Alternatively, P_e could be calculated using the marginal frequencies given in Table 7.11. That is,

$$P_e = \frac{(23)(6) + (43)(48) + (56)(89) + (15)(2)}{(137)(145)}$$

$$= 0.3633$$

$$\kappa = 0.078$$

In this case the two approaches give essentially the same results but the latter strategy would lead to a more efficient use of data in the case of a table such as Table 7.11, particularly in the case where, as here, a variable number of the eight referees assess each of the manuscripts.

Returning to Table 7.13, interest might now be directed at two further aspects of agreement. The first is assessment of agreement for each individual grade and the second is consideration of the combination at grades to potentially improve the level of agreement. Table 7.13(b) summarises the counts

Table 7.11 Ratings of 181 conference abstracts by eight independent referees

1	2	3	4	5	6	7	8		1	2	3	4	5	6	7	8		1	2	3	4	5	6	7	8
A	A	E	A	A	A	A	B		B	C	C	C	B	D	B	A		C	B	E	B	B	D	C	B
A	E	A	A	A	D	B	B		B	B	C	B	C	C	A	C		C	B	B	B	B	D	C	C
B	B	E	A	A	C	A	A		C	B	C	C	A	B	C	B		C	C	C	B	B	E	B	C
A	C	B	A	A	C	B	A		C	A	C	C	B	B	B	C		C	E	C	C	C	B	B	C
B	C	C	A	A	B	A	A		C	B	C	A	C	B	B	C		C	C	B	B	C	C	B	B
B	A	C	A	B	C	A	C		B	A	C	C	C	D	U	B		C	C	B	C	B	C	C	C
B	A	C	A	B	B	A	A		B	E	C	C	C	D	E	A		C	U	C	C	C	B	B	B
A	B	B	B	B	B	A	A		C	B	B	C	C	C	A	B		C	B	C	C	C	B	E	D
B	A	A	E	B	B	A	B		C	B	E	U	D	D	B	B		C	C		U	C	U		
B	B	B	A	B	B	A	A		C	A	E	C	B	C	C	B		C	C	C	E	B	C	B	E
B	A	E	A	A	C	A	B		B	B	C	C	B	B	U	U		C	C	E	C	B	D	E	C
C	A	A	C	B	C	C	A		C	U	C	C	B	D	A	B		C	C	U	C	C	B	C	C
B	B	B	B	C	B	A	A		C	C	B	C	C	U	B	A		C	E	C	C	B	B	E	B
B	A	B	A	B	A	A	B		B	B	B	U	D	D	A	B		C	C	C	C	D	C	B	C
C	B	E	C	A	A	A	A		C	E	C	C	B	E	U	M		C	C	C	U	U	U	B	B
B	A	A	B	B	A	A	B		C	U	C	C	B	C	A	C		C	C	C	C	C	D	B	C
B	B	B	A	B	A	A	A		C	U	U	U	B	B	B	B		U	U	C	U	U	U	E	B
A	A	B	A	A	B	A	U		C	C	C	C	C	U	C	U		C	C	C	C	C	E	M	C
B	B	B	C	C	U	U	B		C	B	C	A	D	U	A	B		C	C	C	C	D	C	M	C
B	A	B	A	B	B	A	A		C	C	C	U	C	D	A	C		B	U	U	U	U	U	B	B
B	B	B	B	C	C	C	B		C	D	B	C	C	D	B	U		C	C	U	B	C	U	B	C
C	U	B	C	C	U	U	C		C	E	U	C	B	E	B	U		U	C	C	U	U	D	B	C
B	B	B	B	B	U	C	C		U	C	U	U	C	C	B	C		U	C	C	C	C	U	B	C
B	A	B	A	B	U	A	B		U	A	M	U	U	U	B	U		C	U	U	U	C	U	U	C
B	B	B	B	C	A	C	B		C	C	U	U	U	D	B	C		C	C	C	B	B	B	C	C
B	U	M	U	B	C	U	C		U	C	W	A	C	A	M	A		B	B	M	C	C	D	C	C
B	B	B	U	C	C	A	C		C	U	C	U	C	C	B	C		C	C	U	B	B	D	U	C
B	A	B	A	B	D	C	B		C	A	U	U	U	U	B	A		U	C	C	U	C	U	C	C
B	C	B	A	C	U	A	B		C	U	C	U	C	D	C	C		C	C	U	M	C	B	U	U
B	A	C	A	C	U	B	B		C	A	D	A	D	A	D	C		C	C	C	U	U	U	U	U
C	A	M	C	B	E	A	B		C	U	E	C	U	U	E	U		U	B	U	U	D	D	C	C
B	C	B	B	B	D	A	B		C	A	C	U	C	U	U	C		C	E	E	U	D	D	B	C
B	U	B	U	U	B	A	B		C	C	U	C	C	A	U	A		C	M	U	U	U	B	B	E
U	D	B	U	C	U	C	B		U	U	E	E	U	U	E	U		U	D	U	D	U	U	U	B
	C	B	U	B	D	A	A		C	D	U	A	U	D	C	A		C	E	D	C	C	D	C	U
B	A	B	U	D	B	C	B		U	A	U	C	U	C	E	C		C	E	E	C	U	C	B	C

```
  C  C B C C C B E B B C B B C C C        C C C D C C C C
B B C C C C E B E C C C E C C B C         C E C C C C E C C C
B D D D D D D C C C D D D C B C C C D D D D C D D D C D D C
C C D C D C D C C C B C C C C B C D C D C D D D C C C C
C D E E C C C C D C C C C C C C C D E C E C C C D C D D C D C
   C C C C B B C C C C C C C C C C        C D C C C C C C C
D E C C D B E C D C D C C C C C C D D D D C E C E E C C C D
C C C C C C C C C C C C B C C C D C C C C C C C C C C C C


C C B B B B B B B B B C B B B E B B B B B C B B B C C B B B
C A C A B B B B B B B B B B B B C C B B B C C C B B B C C
A D B C B D B B B C B D B C D C B D D D B D C C B D E B D B D
C C B B B C B B B B B B C C B B B B B C B B B B C B C B B C C
C C E B B B B B C B B B C B B E D B B C C C C B E B C C B C
E C B B B B B C C B E C C B E B C B B C C C C C E E C E B
E C B B C E E E E E C B B B B E E E E E B B B B B D E E C E
C C C C B C C B C C B C C B C C B C C B B C B B C C C E B


B C A C   B C A C B C C C E E B B B B B B B B B C B
A B A B A A B A A B B A B A A C C A A C C A C B B B C B B
E A B D C B D D B D D B B B C B B B B B D B B D C C C D
B B B C C C B A C B B A C B B B B C B B B A B B B B A D
A B C A B A A A B A B A C C B B B B A C B B B A B A A A
C C E C E E C C B B B B B E B B C A B B C B B B E C C B C
B A C A A C C E E B B B E E B B B B B C B C B D E E B A C B
C C C B C C C C B B B B B B B C B C B B B B B C C C C B C
```

Codes: A Stongly recommended as a spoken contribution to the conference
        B Recommended as a spoken contribution
        C Recommended for acceptance as a short communication on a poster
        D Recommended for rejection
        E The referee has no opinion on this abstract
        blank Not assessed by this referee (missing data)

Table 7.12 Marginal frequencies for the eight referees in Table 7.11

Referee	Coding*					
	A	B	C	D	E	missing
1	6	59	112	1	3	0
2	23	43	56	15	44	0
3	6	48	89	2	27	0
4	33	45	75	10	18	0
5	12	75	74	20	0	0
6	12	54	49	60	6	0
7	41	68	52	1	10	9
8	20	79	66	2	5	9

* See Table 7.11 for definition of codes.

for Referees 2 and 3 when A is combined with B, and C is combined with D. The κ statistic for this table is 0.238. Schouten (1986) provides general conditions that are necessary for the combination of two categories to lead to a higher level of agreement between two raters. Table 7.14 summarizes the results of calculating κ statistics for each of the four grades (A–D) separately. Part (e) of Table 7.14 illustrates how the overall κ of 0.072 can be calculated from the data from the codes examined separately. The overall κ is simply a weighted average of the κ from the individual grades (see Fleiss, 1981).

The extension and generalisations of the κ coefficient to measure agreement between more than two raters has been discussed by several authors (see, for example, Fleiss, 1971; Light, 1971, Hubert, 1977, Conger, 1980a; Davies and Fleiss, 1982; and Schouten, 1986). The approach taken here will be essentially that of Davies and Fleiss (1982). Suppose that each of I objects is classified into C mutually exclusive and exhaustive categories by each of the same set

Table 7.13 Patterns of agreement and disagreement for referees 2 and 3 (see Table 7.11)

(a) Using four codes separately

		Referee 3				
		A	B	C	D	Total
	A	0	9	9	0	18
Referee 2	B	3	13	21	1	38
	C	1	9	32	1	43
	D	0	1	9	0	10
	Total	4	32	71	2	109

(b) After combining codes A and B and also C and D

		Referee 3		
		A and B	C and D	Total
	A and B	25	31	56
Referee 2	C and D	11	42	53
	Total	36	73	109

Table 7.14 Agreement between referees 2 and 3 over use of individual codes

(a) A versus the rest

		Referee 3		
		A	Not A	
Referee 2	A	0	18	$P_o = 0.7982$
	Not A	4	87	$P_e = 0.8103$

(b) B versus the rest

		Referee 3		
		B	Not B	
Referee 2	B	13	25	$P_o = 0.5963$
	Not B	19	52	$P_e = 0.5625$

(c) C versus the rest

		Referee 3		
		C	Not C	
Referee 2	C	32	11	$P_o = 0.5413$
	Not C	39	27	$P_e = 0.4681$

(d) D versus the rest

		Referee 3		
		D	Not D	
Referee 2	D	0	10	$P_o = 0.8899$
	Not D	2	97	$P_e = 0.8933$

(e) Summary table

Grade	$P_o - P_e$	$1 - P_e$	Kappa
A	−0.0121	0.1897	−0.0638
B	0.0338	0.4375	0.0773
C	0.0732	0.5319	0.1376
D	−0.0034	0.1067	−0.0319
Total:	0.0915	1.2658	

$$\text{Overall } \kappa = \frac{0.0915}{1.2658} = 0.072$$

of K observers or raters. The observation for the ith object ($i = 1, \ldots, I$) by the kth rater ($k = 1, \ldots, K$) is represented by the vector

$$\mathbf{X}_{ik} = (X_{ik1}, X_{ik2}, \ldots, X_{ikC})'$$

where each X_{ikc} ($c = 1, \ldots, C$) has the value of 0 or 1. If the rating or grade given is actually c then $X_{ikc} = 1$ but, if not, then $X_{ikc} = 0$. The vector $\mathbf{X}_{ikc}$, therefore, will have only one element with the value of 1, and $C - 1$ other elements with the value 0 (implying that $\sum_{c=1}^{C} X_{ik} = 1$ for all values of both i and k).

Now, for the ith object let

$$Y_{ic} = \sum_{k=1}^{K} X_{ikc}$$

That is, Y_{ic} is the number of raters or observers who allocate the ith object to the rating or category c. It follows that

$$\sum_{c=1}^{C} Y_{ic} = \sum_{k=1}^{K} \sum_{c=1}^{C} X_{ikc} = K$$

For each object, there are a total of $\frac{1}{2}K(K-1)$ pairs of classifications. For the ith object, the observed number of pairs of classifications that are in agreement is

$$\frac{1}{2} \sum_{c=1}^{C} Y_{ic}(Y_{ic} - 1)$$

Adding over the I objects the observed proportion of the possible pairs of classifications that are in agreement is given by

$$P_o = \frac{1}{IK(K-1)} \sum_{i=1}^{I} \sum_{c=1}^{C} Y_{ic}(Y_{ic} - 1)$$

$$= \frac{1}{IK(K-1)} \left(\sum_{i=1}^{I} \sum_{c=1}^{C} Y_{ic}(Y_{ic}^2 - IK) \right) \tag{7.33}$$

Let π_{kc} be the probability that a randomly chosen object is placed into category c by the kth rater or observer. This probability is estimated by

$$p_{kc} = \frac{1}{I} \sum_{i=1}^{I} X_{ikc} \tag{7.34}$$

This is simply the observed proportion of the I objects placed into category c by the kth rater. If the classifications by the raters are statistically independent then the probability that a given pair of raters (k and m, say) will agree in the classification of a randomly selected unit is estimated by

$$\sum_{c=1}^{C} p_{kc} p_{mc}$$

The average chance-expected probability of *pairwise agreement* is then estimated by

$$P_e = \frac{1}{K(K-1)} \sum_{\substack{k=1}}^{K} \sum_{\substack{m=1 \\ m \neq k}}^{K} \sum_{c=1}^{C} p_{kc} p_{mc} \tag{7.35}$$

Finally κ is defined (in an analogous way to that described in Section 2.8) by

$$\kappa = \frac{P_o - P_e}{1 - P_e} \tag{7.36}$$

Equation (7.36) can be shown (Davies and Fleiss, 1982) to be equivalent to

$$\kappa = 1 - \frac{IK^2 - \sum_{i=1}^{I} \sum_{c=1}^{C} Y_{ic}^2}{I\{K(K-1) \sum_{c=1}^{C} \bar{p}_c(1 - \bar{p}_c) + \sum_{c=1}^{C} \sum_{k=1}^{K} (p_{kc} - \bar{p}_c)^2\}} \tag{7.37}$$

where $\bar{p}_c$ is the average (over raters) of the p_{kc}. That is

$$\bar{p}_c = \frac{1}{K} \sum_{k=1}^{K} p_{kc}$$

There are two interesting special cases of Equation (7.37). The first occurs when it can be assumed that the p_{kc} are the same for all values of k (that is, marginal homogeneity). In this case Equation (7.37) reduces to

$$\kappa = 1 - \frac{IK^2 - \sum_{i=1}^{I} \sum_{c=1}^{C} Y_{ic}^2}{I\{K(K-1) \sum_{c=1}^{C} \bar{p}_c(1 - \bar{p}_c)\}} \tag{7.38}$$

This is the coefficient of agreement applicable for the situation where, for each subject, a fixed number of raters are randomly selected from a large pool of potential raters (Fleiss, 1971). It is equivalent to the situation where one derives an intra-class correlation coefficient through the use of a one-way analysis of variance (see Chapter 6).

The second interesting special case is when there are only two possible classifications (that is $c = 1$ or 2). Here Equation (7.37) reduces to

$$\kappa = 1 - \frac{\sum_{i=1}^{I} Y_{i2}(K - Y_{i2})}{I\{K(K-1)\bar{p}_2(1 - \bar{p}_2) + \sum_{k=1}^{K}(p_{k2} - \bar{p}_2)^2\}^2} \tag{7.39}$$

Equation (7.39) is itself equivalent to

$$\kappa = \frac{MSU - MSE}{MSU + (K-1)MSE + (MSO)/(I-1)} \tag{7.40}$$

where MSU, MSO and MSE are mean squares due to objects, raters and measurement error, respectively. These mean squares are derived from a two-way analysis of variance of the binary ratings (see Davies and Fleiss, 1982). The corresponding moments estimator for the intra-class correlation derived from a two-way analysis of variance is given by (see Bartko, 1966)

$$R = \frac{MSU - MSE}{MSU + (K-1)MSE + K(MSO - MSE)/I} \tag{7.41}$$

In practice, the difference between (7.40) and (7.41) is trivial. For the data in Table 7.3, for example, we have already seen that $R = 0.522$ (see Section 7.3). Equation (7.40) yields a κ coefficient of 0.520.

The calculation in Equation (7.37) will be illustrated using the data given in Table 7.15. These data comprise a randomly selected group of 25 of the abstracts listed in Table 7.11. The only restriction imposed in sampling from the complete set of 181 papers is that all of the selected abstracts should have been rated by all of the eight referees (that is, there is no missing data). The results in Table 7.15 are presented in summary form in Table 7.16. This table provides the value of $P_o = 0.3943$. The marginal counts derived from Table 7.15 are presented in Table 7.17. From these one can obtain the value of $P_e = 0.3364$. The value of κ obtained is 0.087.

The standard error of κ as expressed by Equation (7.37) can be calculated using bootstrap or jackknife techniques. However, in the use of these two techniques one has to remember to consider subsampling from both objects *and* raters. Calculation of standard errors of the κ coefficients derived from

Table 7.15 Assessment of 25 abstracts, randomly selected from those in Table 7.11 with complete data

Abstract	Referee							
	1	2	3	4	5	6	7	8
1	C	A	C	C	B	B	B	C
2	B	B	C	B	B	B	C	A
3	B	B	B	B	C	B	A	B
4	C	C	C	A	A	D	B	C
5	C	C	B	C	D	D	C	C
6	B	D	B	B	A	D	C	B
7	C	C	C	C	C	D	C	B
8	C	B	C	A	B	D	B	B
9	C	C	C	D	C	D	C	C
10	B	B	C	C	C	B	B	B
11	C	C	C	C	C	D	B	B
12	C	B	C	C	D	D	C	C
13	C	A	C	C	B	C	C	C
14	C	A	B	B	B	C	C	B
15	B	C	B	C	B	D	A	A
16	B	B	A	A	B	C	A	B
17	C	D	C	A	B	D	C	C
18	C	C	C	C	C	C	B	B
19	C	B	B	B	B	C	B	B
20	C	D	C	C	C	C	C	C
21	C	B	A	B	B	B	C	B
22	B	B	B	A	C	B	A	B
23	C	B	C	C	B	D	B	B
24	B	C	C	C	B	D	B	A
25	C	A	B	B	C	B	A	B

Table 7.15 will be left as an exercise for the reader (see Exercises 7.5). For binary data the large-sample variance of κ can be estimated by (Davies and Fleiss, 1982)

$$\text{Var}(\kappa) = \frac{2[\{\sum_{k=1}^{K} \bar{p}_2(1 - \bar{p}_2)\}^2 - \sum_{k=1}^{K} p_{k2}^2(1 - p_{k2})^2]}{I\{K(K - 1)\bar{p}_2(1 - \bar{p}_2) + \sum_{k=1}^{K}(p_{k2} - \bar{p}_2)^2\}^2} \quad (7.42)$$

Schouten (1986) has considered the use of κ coefficients corresponding to Equation (7.37) when there are missing observations. There are, as before,

$$\frac{1}{2} \sum_{c=1}^{C} Y_{ic}(Y_{ic} - 1)$$

pairs of classifications that are in agreement for the ith subject. If there are K_i observations for the ith subject, then the total number of paired observations is given by $\frac{1}{2}K_i(K_i - 1)$. Adding over the I objects to be classified the observed proportion of the possible pairs of classifications that are in agreement is given by

$$P_o = \frac{1}{I} \sum_{i=1}^{I} \left[\frac{Y_{ic}(Y_{ic} - 1)}{K_i(K_i - 1)} \right] \quad (7.43)$$

Table 7.16 Frequencies of the four codings for the 25 randomly selected abstracts listed in Table 7.15

Abstract	Frequencies (Y_{ic})				$\Sigma_i \Sigma_c Y_{ic}^2$
	A	B	C	D	
1	1	3	4		26
2	1	5	2		30
3	1	6	1		38
4	2	1	4	1	22
5		1	5	2	30
6	1	4	1	2	22
7		1	6	1	38
8	1	4	2	1	22
9			6	2	40
10		5	3		34
11		2	5	1	30
12		1	5	2	30
13	1	1	6		38
14	1	4	3		26
15	2	3	2	1	18
16	3	4	1		26
17	1	1	4	2	22
18		2	6		40
19		6	2		40
20			7	1	50
21	1	5	2		30
22	2	5	1		30
23		4	3	1	26
24	1	3	3	1	20
25	2	4	2		24

Let the subgroup of raters or observers who classify the ith object be represented by G_i. The chance expected proportion of pairwise agreements for object i is given by

$$q_i = \frac{1}{K_i(K_i - 1)} \sum_{k=1}^{K} \sum_{\substack{m=1 \\ m \neq k}}^{K} \sum_{c=1}^{C} p_{kc} p_{mc} \qquad (7.44)$$

where the summation over both k and m is restricted to those raters contained within G_i. From this expression we obtain

$$P_e = \frac{1}{I} \sum_{i=1}^{I} q_i$$

$$= \frac{1}{I} \sum_{i=1}^{I} \frac{1}{K_i(K_i - 1)} \sum_{k=1}^{K} \sum_{\substack{m=1 \\ m \neq k}}^{K} \sum_{c=1}^{C} p_{kc} p_{mc} \qquad (7.45)$$

The marginal proportions, p_{kc} (for all k), are estimated from all of the observations made by observer k. κ is now calculated from P_o and P_e as before.

This section will be completed by pointing out the relationship between κ coefficients defined by Equation (7.37) and the κ coefficients calculated for

Table 7.17 Marginal frequencies derived from Table 7.15

Referee	Classification (rating)				Total
	A	B	C	D	
1	0	8	17	0	25
2	4	10	8	3	25
3	2	8	15	0	25
4	5	7	12	1	25
5	2	12	9	2	25
6	0	7	6	12	25
7	5	9	11	0	25
8	3	14	8	0	25
Total	21	75	86	18	
$\hat{P}_c$	0.105	0.375	0.430	0.090	

each pair of raters separately. For each pair of raters k and m ($k \neq m$) define a κ coefficient as

$$\kappa^{km} = \frac{P_o^{km} - P_e^{km}}{1 - P_e^{km}} \tag{7.46}$$

where P_o^{km} and P_e^{km} are the observed and chance expected proportions of agreement, respectively. These values are given in Table 7.9 for the seven pathologists in Table 7.3. Note from Table 7.9 that, summing over all different raters pairs

$$\frac{\sum (P_o^{km} - P_e^{km})}{\sum (1 - P_e^{km})} = 0.520$$

This is the κ coefficient calculated using Equation (7.37). The latter coefficient is a weighted average of the individual κ, as follows:

$$\kappa = \frac{\sum (1 - P_e^{km})\kappa^{km}}{\sum (1 - P_e^{km})} \tag{7.47}$$

Again the summation is carried out over all different rater pairs (see Schouten, 1985).

7.5   Reliability of unanimous and majority opinions

If no single observation or rating can be demonstrated to have satisfactory reliability (measured using, say, an intra-class correlation or κ coefficient) it still might be possible to use two or more independent observations for each subject or object to give a single consensus or average rating that is reliable enough for clinical or research use. For continuous measurements it has been shown that the reliability of the average of several parallel measurements or test scores can be calculated using the Spearman–Brown prophesy formula (see Equation (3.24)). Kraemer (1979) has shown that for binary measurements the same formula holds approximately – the average measurement being the proportion of the ratings for each subject showing the characteristic in question. If there are K independent raters then the reliability of the proportion of positive ratings is given by

$$R \sim \frac{K\kappa}{1 + K\kappa} \tag{7.48}$$

for small K, and

$$R \sim \frac{K\kappa}{1 + (K - 1)\kappa} \tag{7.49}$$

for large K. Kraemer (1979) left the reader to decide what was meant by a 'large' K. For the seven pathologists in Table 7.3, from (7.49),

$$R \sim \frac{7(0.52)}{1 + 6(0.52)}$$

$$= 0.884 \qquad (0.785 \text{ if } (7.48) \text{ is used})$$

Other authors such as Conger (1980a) and Landis and Koch (1977b) have extended the use of agreement weights for calculation of weighted κ coefficients to measure the degree of simultaneous or majority agreement. The interpretation of these agreement statistics is not, however, particularly easy and the discussion here will move in a different direction. This chapter will end as it started – by consideration of the effectiveness (sensitivity and specificity) of diagnostic tests or screening questionnaires. The argument essentially follows that of Lachenbruch (1988).

Consider, for example, a setting in which K different diagnostic tests are used on each subject. Consider two rules for the combination of the individual test results. To declare a subject positive (suffering from AIDS or psychiatric illness, for example), the *unanimity rule* requires that all of the individual tests yield positive results. The *majority rule* requires that a majority of the individual tests yields positive results. For the pathology example given in Table 7.3 the unanimity rule requires that all seven pathologists recognise carcinoma in a given biopsy slide and the majority rule requires that at least four out of the seven recognise carcinoma in a slide. In general, the unanimity rule can be used for any number of tests while the majority rule, of course, is dependent on there being an odd number of diagnostic tests being used.

Lachenbruch (1988) discusses a sequential strategy for the use of the K tests. The tests are assumed to be given in a fixed order and the second test is applied after the results of the first are known, the third test is applied after the results of the second are known, and so on. This allows the use of sequential stopping rules to lower the cost of the testing process. Using this sequential approach together with the unanimity rule, for example, implies that the testing ceases at the first negative result or after K tests, whichever comes earlier.

Consider the use of three diagnostic tests. If the tests are given in a fixed order the unanimity rule implies that the negative results are $-$ (tests 2 and 3 are not given), $+-$ (test 3 is not given), and $++-$. Assuming, for simplicity, that each of the three tests have the same specificity (s) and also the same sensitivity (r), then the specificity of the unanimous verdict is given by

$$V = s + (1 - s)s + (1 - s)^2 s$$

$$= 1 - (1 - s)^3 \tag{7.50}$$

Similarly the sensitivity of the unanimous verdict is given by

$$U = r^3 \qquad (7.51)$$

The sensitivity of the unanimity rule is always less than that of the individual tests. The specificity of the unanimity rule, on the other hand, is always greater than for an individual test.

The majority rule strategy for three diagnostic tests will declare a subject positive if the test results are $++$ (test 3 is not given), $+-+$ or $-++$, and negative if the results are $--$ (test 3 is not given), $-+-$ or $+--$. The specificity, V, is now given by

$$V = s^2(1 + 2(1 - s)) \qquad (7.52)$$

Similarly, the sensitivity, U, is given by

$$U = r^2(1 + 2(1 - r)) \qquad (7.53)$$

Lachenbruch (1988) also considers the use of two different tests; an inexpensive one with a sensitivity of r_I and specificity of s_I; and an expensive one with sensitivity of r_E and specificity of s_E. If the unanimity rule is used for the two tests then

$$U = r_I r_E \qquad (7.54)$$

and

$$V = s_I + (1 - s_I)s_E \qquad (7.55)$$

Note that it is cheaper to perform the inexpensive test first. For the majority rule strategy two inexpensive tests are first performed and then the expensive one is performed only if the two inexpensive ones disagree. Then for this procedure

$$U = r_I^2 + 2(1 - r_I)r_I r_E \qquad (7.56)$$

and

$$V = s_I^2 + 2(1 - s_I)s_I s_E \qquad (7.57)$$

The above derivations are based on the assumption that the sensitivities and specificities of the tests are constant across subjects. Lachenbruch (1988) also considers cases where this assumption is relaxed. Note also that the above derivations are dependent on the assumption of independence of the diagnostic tests (see Section 7.1).

7.6 Exercises

7.1 Examine the data in Table 7.11 for patterns of bias.

7.2 Recode the data in Table 7.11 to 1 representing either A or B and 0 representing either C or D (regard E as a missing value). Examine patterns of bias through the use of Cochran's Q-statistic and McNemar's X^2-statistic. Examine patterns of agreement or reliability through the use of the various ANOVA approaches described in Section 7.3.

7.3 Treat the codes A, B, C and D in Table 7.11 as representing an ordinal scale. Examine patterns of agreement between pairs of referees through the use of weighted κ coefficients (using linear or quadratic disagreement weights). Compare the results with the use of ANOVA techniques for this ordinal measurement scale.

7.4   Estimate weighted κ coefficients for the two-way agreements/disagreements presented in Table 7.13. Use the jackknife method to estimate the standard errors of these coefficients. (*Hint:* there are only as many unique pseudovalues as there are non-empty cells in the two-way table of counts.)

7.5   Use bootstrap and jackknife techniques to investigate the precision of overall κ coefficients of agreement of the data presented in Table 7.15.

7.6   Investigate patterns of unanimous and majority agreement for the data given in Tables 7.3, 7.11 or 7.15.

Appendix 1

Expected Values

First consider measurements that can have only certain discrete values. A typical example is the number of correct answers to a series of questions designed to test a subject's ability at mathematics. Another example is a count of the number of red blood cells in a small volume of blood suitably diluted in a salts solution. The possible values for these two measurements would all be integers, but this is not always the case. The first score could be converted to the corresponding proportion of correct answers and it would still only take particular discrete values. Let this discrete measurement be represented by the random variable X. Let the possible values of X that can be observed range from x_1 through to x_k (with a typical value of, say, x_i). Further, let the probability that a value x_i is observed be represented by $\text{Prob}(X = x_i)$ or, more simply, by $P(x_i)$. The sequence of probabilities $P(x_1), \ldots, P(x_i), \ldots, P(x_k)$ defines the probability distribution of the random variable X.

The *expected value* or mean of this random variable is defined by

$$E(X) = x_1 P(x_i) + x_2 P(x_2) + \cdots + x_i P(x_i) + \cdots + x_k P(x_k)$$

$$= \sum_{i=1}^{k} x_i P(x_i) \tag{A1.1}$$

If we denote the mean of X by μ_x, then the *variance* of X is defined by

$$\text{Var}(X) = E(X - \mu_x)^2 \tag{A1.2}$$

where the expected value of any function of X, $F(X)$, in this case $(X - \mu_x)^2$, is defined by

$$E[F(X)] = \sum_{i=1}^{k} F(x_i)P(x_i) \tag{A1.3}$$

Now consider a measurement that is continuous (potentially at least, if not in practice). Typical examples would be measurements of weight, time or length. If a continuous measurement is represented by the random variable X, then the probability of observing a value of X between x and $x + dx$, where dx is an infinitesimally small change in X, is denoted by $p(x) \, dx$. The term $p(x)$ is the *probability density function* (or p.d.f.) of the random variable X.

In this case the expected value of X is defined by integrating over all of the possible values of X, that is

$$E(X) = \int_x p(x)\,dx \qquad (A1.4)$$

The variance of X is defined as in Equation (A1.2), but in this case Equation (A1.3) is replaced by

$$E[F(X)] = \int_x F(x)p(x)\,dx \qquad (A1.5)$$

In the manipulation of expected values the results are the same for both discrete and for continuous measurements. In the discussion below these two cases will not therefore be distinguished. Equation (A1.2) can be expressed in a slightly different form:

$$\begin{aligned}
\text{Var}(X) &= E(X - \mu_x)^2 \\
&= E(X^2 - 2X\mu_x + \mu_x^2) \\
&= E(X^2) - 2\mu_x E(X) + \mu_x^2 \\
&= E(X^2) - \mu_x^2 \qquad (A1.6)
\end{aligned}$$

If the reasoning behind this derivation is not intuitively clear to the reader, then it is suggested that he or she demonstrates, through the use of one or two simple examples, that the two expressions for the variance of X (that is, (A1.2) and (A1.6) are equivalent). For an alternative discussion see Bulmer (1979).

The *standard deviation* of X (or, equivalently, the *standard error* of X) is the square root of the variance of X. The ratio obtained by dividing the standard deviation by the corresponding mean is defined as the *coefficient of variation* of a measurement, $C(X)$:

$$C(X) = \sqrt{\frac{\text{Var}(X)}{\mu_x^2}} \qquad (A1.7)$$

Quite often one is interested in the way in which two types of measurement, such as a subject's height and weight, co-vary. If these two measurements are represented by two random variables, X_1 and X_2, with expected values μ_1 and μ_2 respectively, then the *covariance* of X_1 and X_2 can be defined by the following expression:

$$\text{Cov}(X_1, X_2) = E[(X_1 - \mu_1)(X_2 - \mu_2)] \qquad (A1.8)$$

$$= E(X_1 X_2) - \mu_2 E(X_1) - \mu_1 E(X_2) + \mu_1 \mu_2$$

$$= E(X_1 X_2) - \mu_1 \mu_2 \qquad (A1.9)$$

It should be clear to the reader that the covariance of a random variable with itself is the same as its variance. If the standard deviations of X_1 and X_2 are denoted by $\text{sd}(X_1)$ and $\text{sd}(X_2)$, respectively, then the *correlation* (or more fully, the product-moment correlation) between X_1 and X_2 is defined by

$$\text{Corr}(X_1, X_2) = \frac{\text{Cov}(X_1, X_2)}{\text{sd}(X_1)\text{sd}(X_2)} \qquad (A1.10)$$

Exercises

A1.1   Show that the mean or expected value of a discrete random variable, X, known to be distributed as a Poisson variate, is λ, where

$$P(x_i) = \frac{\lambda^{x_i} e^{-\lambda}}{x_i!}$$

Show that the variance of X is also λ.

A2.2   Show that the mean and variance of X, the number of successes from n Bernoulli trials with probability of each success being p, are np and $np(1 - p)$, respectively. Note that

$$P(x_i) = \binom{n}{x_i} p^{x_i}(1 - p)^{n-x_i}$$

A1.3   Consider repeated throws of a six-sided die. At each throw one can observe a number (integer) between 1 and 6, inclusive. Assuming that the die is fair (each side has the same probability of being observed on each throw), what are the expected value and variance of the observations?

A1.4   Suppose that an urn contains N balls of which R are red and the rest, $N - R$, are black. Let n balls be drawn at random, and without replacement, from the urn. Let a random variable X_i be equal to 1 if the ith ball drawn from the urn is red, 0 otherwise. The probability that the ith ball is red is R/N and the probability that both the ith and jth balls are red is $R(R - 1)/N(N - 1)$. Find $E(X_i)$, $\mathrm{Var}(X_i)$ and $\mathrm{Cov}(X_i, X_j)$. Note that in the discrete case, Equation (A1.8) is equivalent to

$$\mathrm{Cov}(X_i, X_j) = \sum_{x_a} \sum_{x_b} [\mathrm{Prob}(X_i = x_a \text{ and } X_j = x_b)(x_a - E(X_i))(x_b - E(X_j))]$$

A1.5   Consider the allocation of n subjects to k mutually exclusive diagnostic categories. The joint probability of finding x_1 subjects in category 1, x_2 in category 2, ..., x_k in category k, is given by the multinomial distribution

$$p(x_1, x_2, \ldots, x_k) = \frac{n!}{x_1! x_2! \cdots x_k!} \pi_1^{x_1}, \pi_2^{x_2}, \ldots, \pi_k^{x_k}$$

$$= n! \prod_{i=1}^{k} \frac{\pi_i^{x_i}}{x_i!}$$

where π_i is the probability of being allocated to category i ($i = 1, 2, \ldots, k$). Show that $E(x_i) = n\pi_i$, $\mathrm{Var}(x_i) = n\pi_i(1 - \pi_i)$ and $\mathrm{Cov}\,(x_i, x_j) = -n\pi_i\pi_j$ for all i, j.

Appendix 2

Linear Combinations of Measurements

Quite often one is interested in the precision or reliability of, for example, a simple change score or difference, an arithmetic mean of several item scores, or a weighted average or combination of item scores. All of these combinations of the original measurements or item scores can be regarded as *linear combinations* of these scores or as measures derived by a linear transformation of these original scores. A linear combination, Y, of k random variates X_1, $X_2, \ldots, X_k$ has the following general form:

$$Y = a_0 + a_1 X_1 + a_2 X_2 + \cdots + a_k X_k \tag{A2.1}$$

The terms $a_0, a_1, a_2, \ldots, a_k$ are constant coefficients which can take any real value (positive or negative). Examples of linear combinations are the following

$$Y = X_1 - X_2 \tag{A2.2}$$

$$Y = 2 + 3X_1 \tag{A2.3}$$

$$Y = \frac{X_1}{k} + \frac{X_2}{k} + \frac{X_3}{k} + \cdots + \frac{X_k}{k} \tag{A2.4}$$

$$Y = \frac{\sum_{i=1}^{k} w_i X_i}{\sum_{i=1}^{k} w_i} \tag{A2.5}$$

Y in Equation (A2.2) is a simple difference score ($a_0 = 0$, $a_1 = 1$ and $a_2 = -1$). In Equation (A2.3) Y simply reflects a change of scale (adding 2 to every X_1 after first expanding the scale by a factor of 3). A simple arithmetic mean of the k random variables is given in Equation (A2.4) – in this case the X_i could be measurements of different types from a simple individual (including repeated measures of the same trait) or measurements of the same type made on k different individuals. Equation (A2.5) defines a weighted average of the X_i with weights $w_i / \sum w_i$. Finally, (A2.4) is a special case of (A2.5) where $w_i = 1/k$ for all i ($\sum_i w_i = 1$). In the discussion below the term a_0 will usually be ignored and the main aim will be to derive expected values and variances of and covariances between, linear combinations of the following form

$$Y_s = a_{s1}X_1 + a_{s2}X_2 + \cdots + a_{si}X_i + \cdots + a_{sk}X_k \qquad (A2.6)$$

where a_{si} is the constant coefficient for transformation s and random variable X_i. Equation (A2.6) can be also written in terms of the vectors $\mathbf{X}' = (X_1, X_2, \ldots, X_k)$ and $\mathbf{a}'_s = (a_{s1}, a_{s2}, \ldots, a_{sk})$ as follows

$$\mathbf{Y}_s = \mathbf{a}'_s \mathbf{X} \qquad (A2.7)$$

For a vector of transformations $\mathbf{Y}' = (Y_1, Y_2, \ldots, Y_u)$ we have the following matrix expression

$$\mathbf{Y} = \mathbf{AX} \qquad (A2.8)$$

where the matrix $\mathbf{A}$ is given by

$$
\mathbf{A} =
\begin{bmatrix}
a_{11} & a_{12} & \cdots & a_{1i} & \cdots & a_{1k} \\
a_{21} & a_{22} & \cdots & a_{2i} & \cdots & a_{2k} \\
\vdots & & & & & \\
a_{s1} & a_{s2} & \cdots & a_{si} & \cdots & a_{sk} \\
\vdots & & & & & \\
a_{u1} & a_{u2} & \cdots & a_{ui} & \cdots & a_{uk}
\end{bmatrix}
\qquad (A2.9)
$$

Before considering the derivation of the covariance matrix of transformations of the type given in Equation (A2.6), however, a simple linear transformation of a single measurement, X_1, will be considered. Let

$$Y = a_0 + a_1 X_1 \qquad (A2.10)$$

$$E(Y) = a_0 + a_1 E(X_1)$$

$$= a_0 + a_1 \mu_1 \qquad (A2.11)$$

where $\mu_1 = E(X_1)$. In addition,

$$\begin{aligned}
\mathrm{Var}(a_0 + a_1 X_1) &= E[(a_0 + a_1 X_1) - (a_0 + a_1 \mu_1)]^2 \\
&= E[a_1(X_1 - \mu_1)]^2 \\
&= a_1^2 E(X_1 - \mu_1)^2 \\
&= a_1^2 \, \mathrm{Var}(X_1) \qquad (A2.12)
\end{aligned}$$

Now consider a pair of measurements, X_1 and X_2, with means μ_1 and μ_2, respectively.

Let

$$Y = X_1 + X_2$$

Then

$$\begin{aligned}
E(Y) &= E(X_1 + X_2) \\
&= E(X_1) + E(X_2) \\
&= \mu_1 + \mu_2
\end{aligned}$$

Also,

$$\begin{aligned}
\text{Var}(Y) &= E[(X_1 + X_2) - (\mu_1 + \mu_2)]^2 \\
&= E[(X_1 - \mu_1) + (X_2 - \mu_2)]^2 \\
&= E[(X_1 - \mu_1)^2 + (X_2 - \mu_2)^2 + 2(X_1 - \mu_1)(X_2 - \mu_2)] \\
&= E[(X_1 - \mu_1)^2 + E(X_2 - \mu_2)^2] + 2E[(X_1 - \mu_1)(X_2 - \mu_2)]
\end{aligned}$$

$$\text{Var}(X_1 + X_2) = \text{Var}(X_1) + \text{Var}(X_2) + 2\,\text{Cov}(X_1, X_2) \tag{A2.13}$$

In particular, if X_1 and X_2 are statistically independent, then

$$\text{Var}(X_1 + X_2) = \text{Var}(X_1) + \text{Var}(X_2) \tag{A2.14}$$

Similarly,

$$\text{Var}(X_1 - X_2) = \text{Var}(X_1) + \text{Var}(X_2) - 2\,\text{Cov}(X_1, X_2) \tag{A2.15}$$

And, again, if X_1 and X_2 are statistically independent, then

$$\text{Var}(X_1 - X_2) = \text{Var}(X_1 + X_2) \tag{A2.16}$$

It is quite straightforward to use the same arguments to show that

$$E\left(\sum_{i=1}^{k} X_i\right) = \sum_{i=1}^{k} \mu_i \tag{A2.17}$$

where $E(X_i) = \mu_i$ for $i = 1, 2, \ldots, k$.

$$\text{Var}\left(\sum_{i=1}^{k} X_i\right) = \sum_{i=1}^{k} \text{Var}(X_i) + 2 \sum_{i=1}^{k} \sum_{\substack{j=1 \\ j>i}}^{k} \text{Cov}(X_i, X_j) \tag{A2.18}$$

If $\text{Cov}(X_i, X_j) = 0$ for all i and j ($i \neq j$), then

$$\text{Var}\left(\sum_{i=1}^{k} X_i\right) = \sum_{i=1}^{k} \text{Var}(X_i) \tag{A2.19}$$

and, in addition, if the measurements, X_i, all have the same variance, say $\text{Var}(X)$, then

$$\text{Var}(\bar{X}) = \text{Var}\left(\frac{\sum_{i=1}^{k} X_i}{k}\right) = \frac{1}{k}\text{Var}\left(\sum_{i=1}^{k} X_i\right)$$

$$= \frac{1}{k}\text{Var}(X) \tag{A2.20}$$

Now consider the linear combination given in (A2.6):

$$E(Y_s) = \sum_{i=1}^{k} a_{si}\mu_i \tag{A2.21}$$

where $\mu_i = E(X_i)$ as before.

$$\text{Var}(Y_s) = \sum_{i=1}^{k} a_{si}^2 \text{Var}(X_i) + 2 \sum_{i=1}^{k} \sum_{\substack{j=1 \\ j>i}}^{k} a_{si}a_{sj}\text{Cov}(X_i, X_j) \tag{A2.22}$$

This can be represented in matrix terms by the equation

$$\text{Var}(Y_s) = \mathbf{a}_s' \mathbf{V_X} \mathbf{a}_s \tag{A2.23}$$

where the matrix $\mathbf{V_X}$ is the covariance matrix of $\mathbf{X}' = (X_1, X_2, \ldots, X_k)$. That is,

$$\mathbf{V_X} = \begin{bmatrix} \text{Var}(X_1) & \text{Cov}(X_1, X_2) & \cdots & \text{Cov}(X_1, X_k) \\ \text{Cov}(X_2, X_1) & \text{Var}(X_2) & \cdots & \text{Cov}(X_2, X_k) \\ \vdots & \vdots & & \vdots \\ \text{Cov}(X_k, X_1) & \text{Cov}(X_k, X_2) & \cdots & \text{Var}(X_k) \end{bmatrix} \tag{A2.24}$$

The information given in Equation (A2.22) can be used to calculate the variance of the weighted average described by (A2.5):

$$\text{Var}(Y) = \frac{1}{(\sum_{i=1}^k w_i)^2} \text{Var}\left(\sum_{i=1}^k w_i X_i \right)$$

$$= \frac{1}{(\sum_{i=1}^k w_i)^2} \left[\sum_{i=1}^k w_i^2 \text{Var}(X_i) + 2 \sum_{i=1}^k \sum_{\substack{j=1 \\ j>i}}^k w_i w_j \text{Cov}(X_i, X_j) \right] \tag{A2.25}$$

Now, if the measurements are all independent of each other and weights are chosen such that $w_i = 1/\text{Var}(X_i)$, then

$$\text{Var}(Y) = \frac{1}{(\sum_{i=1}^k w_i)^2} \left[\sum_{i=1}^k w_i \right]$$

$$= \frac{1}{\sum_{i=1}^k w_i} \tag{A2.26}$$

Finally, all that is needed now is to derive the variance covariance matrix of $\mathbf{Y}$ given by Equation (A2.8). Consider two linear combinations Y_s and Y_t, where

$$Y_s = a_{s1} X_1 + a_{s2} X_2 + \cdots + a_{si} X_i + \cdots + a_{sk} X_k$$

and

$$Y_t = a_{t1} X_1 + a_{t2} X_2 + \cdots + a_{ti} X_i + \cdots + a_{tk} X_k$$

The variances of Y_s and Y_t have a similar form to Equations (A2.22) and (A2.23). The covariance of Y_s and Y_t is given by

$$\begin{aligned} \text{Cov}(Y_s, Y_t) &= E[(Y_s - E(Y_s))(Y_t - E(Y_t))] \\ &= E[Y_s Y_t - Y_s E(Y_t) - Y_t E(Y_s) + E(Y_s)E(Y_t)] \\ &= E[Y_s Y_t] - E(Y_s)E(Y_t) \\ &= E\left[\sum_{i=1}^k \sum_{j=1}^k a_{si} a_{tj} X_i X_j \right] - \sum_{i=1}^k \sum_{j=1}^k a_{si} a_{tj} \mu_i \mu_j \\ &= \sum_{i=1}^k \sum_{j=1}^k a_{si} a_{tj} E[X_i X_j - \mu_i \mu_j] \\ &= \sum_{i=1}^k \sum_{j=1}^k a_{si} a_{tj} \text{Cov}(X_i, X_j) \end{aligned} \tag{A2.27}$$

In matrix terms, this is equivalent to

$$\text{Cov}(Y_s, Y_t) = \mathbf{a}_s' \mathbf{V_X} \mathbf{a}_t \tag{A2.28}$$

For U linear combinations, $\mathbf{Y}$, the covariance matrix of $\mathbf{Y}$ is given by

$$\mathbf{V_Y} = \mathbf{A} \mathbf{V_X} \mathbf{A}' \tag{A2.29}$$

where the matrix of coefficients, $\mathbf{A}$, is given in (A2.9).

Exercises

A2.1   The random variables X_1 and X_2 have means of 3 and 5, respectively. Their covariance is -2. What are the variances of the following?

(a) $X_1 + X_2$ (b) $X_1 - 2X_2$ (c) $\dfrac{X_1 + 3X_2}{2}$ (d) $6 + 3X_2 - 4X_1$.

A2.2   (a) The variance of X_1 is 2 and that of X_2 is 4 and their covariance is 1. What is the covariance of $X_1 - X_2$ and $X_1 + X_2$ (i.e. $\text{Cov}(X_1 - X_2, X_1 + X_2)$)?

(b) The variance of both X_1 and of X_2 is equal to 2. Their covariance is 1. What is $\text{Cov}(X_1 - X_2, X_1 + X_2)$?

A2.3   Assume that the counts in a 3×3 contingency table are distributed according to the multinomial distribution (see Exercise A1.5). What is the covariance matrix for the six marginal totals in terms of the total count (N) and the cell probabilities (π_{ij})?

A2.4   The variance equations derived in this Appendix apply to expected values. They also apply to sample estimates. Demonstrate that this is true for Equations (A2.13) and (A2.15).

Appendix 3

The δ Technique

The δ method is a technique whereby one uses the Taylor series expansion for a function of one or more random variables to obtain approximations to the expected value of the function and to its variance. In more complex cases it is also used to approximate the covariance matrix of a vector of functions. Consider, for example, a situation in which a physics student wishes to estimate the volume of a right circular cylinder (Pugh and Winslow, 1966, Chapter 11). The volume, V, of a cylinder is given by the equation

$$V = \tfrac{1}{4}\pi d^2 h \tag{A3.1}$$

where d is the diameter of the cylinder and h is its height. The student, through the use of repeated measurements, has estimated d and h together with their standard errors of measurement, σ_d and σ_h, respectively. The student's problem is to obtain a standard error for the estimate of the cylinder's volume.

A second elementary example from physics concerns the measurement of g, the acceleration due to gravity, using a simple pendulum (Taylor, 1982, Section 3.7). The period of a simple pendulum is known to be

$$T = 2\pi \sqrt{\frac{l}{g}} \tag{A3.2}$$

where l is the length of the pendulum. Typically a physics student will be asked to measure the length of the pendulum, l, and its corresponding period, T. Then g can be estimated using the rearrangement of (A3.2).

$$g = \frac{4\pi^2 l}{T^2} \tag{A3.2}$$

Again, on the assumption that the student has used repeated measurements of l and T to estimate their standard errors of measurement, σ_l and σ_T, respectively, the problem is to estimate the standard error of g. These two examples illustrate what physicists call the *propagation of errors*. Similar problems, however, are found in many other areas other than in the physical sciences.

Before the estimation of σ_V and σ_g for Equations (A3.1) and (A3.3) is considered, the general case of a function of a single random variable or measurement, X, will be discussed. What are the expected value and variance of $F(X)$, where $F(X)$ is a differentiable function of X? Let

$$X = \mu + e \tag{A3.4}$$

where $E(X) = \mu$ and $E(e) = 0$, and $\mathrm{Var}(X) = E(e^2)$. The e term is assumed to be small relative to μ. Then

$$F(X) = F(\mu + e) \tag{A3.5}$$

which, following the Taylor expansion, is equivalent to

$$F(X) = \sum_{k=0}^{\infty} \frac{(X - \mu)^k}{k!} \frac{\mathrm{d}^k(F(X))}{\mathrm{d}X^k} \bigg|_{X = \mu} \tag{A3.6}$$

$$= F(\mu) + e \frac{\mathrm{d}(F(X))}{\mathrm{d}X} \bigg|_{X = \mu} + \frac{e^2}{2} \frac{\mathrm{d}^2(F(X))}{\mathrm{d}X^2} \bigg|_{X = \mu} + \cdots \tag{A3.7}$$

If terms involving e^2, e^3, etc. are assumed to be negligible, then

$$F(X) \sim F(\mu) + e \frac{\mathrm{d}(F(X))}{\mathrm{d}X} \bigg|_{X = \mu} \tag{A3.8}$$

If the second form of this expression is represented by $eF'(\mu)$, then

$$F(X) \sim F(\mu) + eF'(\mu) \tag{A3.9}$$

$F'(\mu)$ is the first derivative of $F(X)$ with respect to X, evaluated at $X = \mu$. Readers unfamiliar with the Taylor series expression are referred to a book on calculus such as Apostol (1967).

From (A3.9)

$$E[F(X)] \sim E[F(\mu) + eF'(\mu)]$$

$$= F(\mu) + F'(\mu)E(e)$$

$$= F(\mu) \tag{A3.10}$$

and

$$\mathrm{Var}[F(X)] = E[F(X) - E[F(X)]]^2$$

$$\sim E[F(X) - F(\mu)]^2$$

$$= E[F(\mu) + eF'(\mu) - F(\mu)]^2$$

$$= E[eF'(\mu)]^2$$

$$= [F'(\mu)]^2 E(e^2)$$

$$= [F'(\mu)]^2 \, \mathrm{Var}(X) \tag{A3.11}$$

In practice μ will be unknown and therefore will have to be replaced by its estimate ($\hat{\mu}$) in this equation.

The use of Equation (A3.11) will now be illustrated through four simple examples of $F(\ )$. These are $Y = a_0 + a_1 X$ (see Appendix 2), $Y = \log_e X$, $Y = \exp(X)$, and $Y = 1/X$. First consider $Y = a_0 + a_1 X$.

In this case

$$F'(\mu) = a_1 \tag{A3.12}$$

and

$$\mathrm{Var}(a_0 + a_1 X) \sim a_1^2\,\mathrm{Var}(X) \tag{A3.13}$$

This is identical to the exact result derived in Appendix 2. Now consider the second example, $Y = \log_e X$. Here

$$F'(\mu) = \frac{1}{\mu} \tag{A3.14}$$

and therefore

$$\mathrm{Var}[\log_e(X)] \sim \frac{1}{\mu^2}\,\mathrm{Var}(X) \tag{A3.15}$$

For the third example, $Y = \exp(X)$ or e^X

$$F'(\mu) = \exp(\mu) = e^\mu \tag{A3.16}$$

and

$$\mathrm{Var}(e^X) \sim e^{2\mu}\,\mathrm{Var}(X) \tag{A3.17}$$

Finally, for the reciprocal,

$$F'(\mu) = -\frac{1}{\mu^2} \tag{A3.18}$$

and

$$\mathrm{Var}\!\left(\frac{1}{X}\right) \sim \frac{1}{\mu^4}\,\mathrm{Var}(X) \tag{A3.19}$$

The case of a function, $F(\ )$, of two random variables, X_1 and X_2 is treated in a similar fashion. Let $X_1 = \mu_1 + e_1$ and $X_2 = \mu_2 + e_2$ where $E(X_1) = \mu_1$, $E(X_2) = \mu_2$ and $E(e_1) = E(e_2) = 0$. In addition, $\mathrm{Var}(X_1) = E(e_1^2)$, $\mathrm{Var}(X_2) = E(e_2^2)$ and $\mathrm{Cov}(X_1, X_2) = E(e_1 e_2)$. Using a bivariate Taylor series expansion and a similar approximation to that used in Equation (A3.8) one obtains

$$F(X_1, X_2) = F(\mu_1 + e_1, \mu_2 + e_2)$$

$$\sim F(\mu_1, \mu_2) + e_1 \left.\frac{\partial(F(X_1, X_2))}{\partial X_1}\right|_{\substack{X_1 = \mu_1 \\ X_2 = \mu_2}} \tag{A3.20}$$

$$+ e_2 \left.\frac{\partial(F(X_1, X_2))}{\partial X_2}\right|_{\substack{X_1 = \mu_1 \\ X_2 = \mu_2}}$$

$$= F(\mu_1, \mu_2) + e_1 \frac{\partial F(\ )}{\partial X_1} + e_2 \frac{\partial F(\ )}{\partial X_2} \tag{A3.21}$$

where $\partial F(\ )/\partial X_1$ and $\partial F(\ )/\partial X_2$ represent the first partial derivatives of $F(X_1, X_2)$ with respect to X_1 and X_2, respectively, again evaluated at $X_1 = \mu_1$ and $X_2 = \mu_2$. If μ_1 and μ_2 are not known they can be replaced by their corresponding estimates:

$$E(F(X_1, X_2)) \sim E\left[F(\mu_1, \mu_2) + e_1 \frac{\partial F(\ )}{\partial X_1} + e_2 \frac{\partial F(\ )}{\partial X_2}\right]$$

$$= F(\mu_1, \mu_2) \tag{A3.22}$$

$$\text{Var}[F(X_1, X_2)] \sim E[F(X_1, X_2) - F(\mu_1, \mu_2)]^2$$

$$= E\left[F(\mu_1, \mu_2) + e_1 \frac{\partial F(\ )}{\partial X_1} + e_2 \frac{\partial F(\ )}{\partial X_2} - F(\mu_1, \mu_2)\right]^2$$

$$= E\left[e_1 \frac{\partial F(\ )}{\partial X_1} + e_2 \frac{\partial F(\ )}{\partial X_2}\right]^2$$

$$= E\left[e_1^2\left(\frac{\partial F(\ )}{\partial X_1}\right)^2 + 2e_1 e_2 \frac{\partial F(\ )}{\partial X_1}\frac{\partial F(\ )}{\partial X_2} + e_2^2\left(\frac{\partial F(\ )}{\partial X_2}\right)^2\right]$$

$$= \left(\frac{\partial F(\ )}{\partial X_1}\right)^2 E(e_1^2) + 2\frac{\partial F(\ )}{\partial X_1}\frac{\partial F(\ )}{\partial X_2} E(e_1 e_2) + \left(\frac{\partial F(\ )}{\partial X_2}\right)^2 E(e_2^2)$$

$$= \left(\frac{\partial F(\ )}{\partial X_1}\right)^2 \text{Var}(X_1) + 2\frac{\partial F(\ )}{\partial X_1}\frac{\partial F(\ )}{\partial X_2} \text{Cov}(X_1, X_2)$$

$$+ \left(\frac{\partial F(\ )}{\partial X_2}\right)^2 \text{Var}(X_2) \tag{A3.23}$$

The use of this equation will now be illustrated through reference to the two physical examples given at the beginning of this Appendix. First consider the volume, V, of the right circular cylinder. From Equation (A3.1)

$$\frac{\partial V}{\partial d} = \frac{\pi}{2} dh \tag{A3.24}$$

and

$$\frac{\partial V}{\partial h} = \frac{\pi}{4} d^2 \tag{A3.25}$$

If one now assumes that $\text{Cov}(d, h) = 0$, then

$$\sigma_V^2 \sim \left(\frac{\partial V}{\partial d}\right)^2 \sigma_d^2 + \left(\frac{\partial V}{\partial h}\right)^2 \sigma_h^2$$

$$= \left(\frac{\pi^2}{4} d^2 h^2\right)\sigma_d^2 + \left(\frac{\pi^2}{16} d^4\right)\sigma_h^2 \tag{A3.26}$$

The physics student may, however, be more interested in the coefficient of variation of V (that is, in fractional errors). This can be obtained through division of Equation (A3.26) throughout by V^2 to give

$$\frac{\sigma_V^2}{V^2} \sim \frac{4\sigma_d^2}{d^2} + \frac{\sigma_h^2}{h^2} \tag{A3.27}$$

The coefficient of variation of V is given by the square root of the right-hand side of (A3.27).

For the pendulum example, from Equation (A3.3),

$$\frac{\partial g}{\partial l} = \frac{4\pi^2}{T^2} \tag{A3.28}$$

$$\frac{\partial g}{\partial T} = -\frac{8\pi^2 l}{T^3} \tag{A3.29}$$

and, again assuming that $Cov(l, T) = 0$,

$$\sigma_g^2 \sim \left(\frac{16\pi^4}{T^4}\right)\sigma_l^2 + \left(\frac{64\pi^4 l^2}{T^6}\right)\sigma_T^2 \tag{A3.30}$$

The coefficient of variation of g is given by

$$\sqrt{\left(\frac{\sigma_g^2}{g^2}\right)} \sim \sqrt{\left(\frac{\sigma_l^2}{l^2} + \frac{4\sigma_T^2}{T^2}\right)} \tag{A3.31}$$

Consider one further example, that is, the product of two independent random variables, $Y = X_1 X_2$. In this case it is possible to derive an approximate expression of the variance of the product and it is also straightforward to produce an exact formula. If $E(x_1) = \mu_1$ and $E(X_2) = \mu_2$ as before, then, in the case of independence

$$E(X_1 X_2) = E(X_1)E(X_2)$$

$$= \mu_1 \mu_2$$

it follows that

$$\begin{aligned} Var(X_1 X_2) &= E[X_1 X_2 - E(X_1 X_2)]^2 \\ &= E(X_1^2 X_2^2) - [E(X_1 X_2)]^2 \\ &= E(X_1^2)E(X_2^2) - [E(X_1)E(X_2)]^2 \\ &= [Var(X_1) + \mu_1^2][Var(X_2) + \mu_2^2] - \mu_1^2 \mu_2^2 \\ &= Var(X_1)\,Var(X_2) + \mu_2^2\,Var(X_1) + \mu_1^2\,Var(X_2) \quad (A3.32) \end{aligned}$$

It will be left as an exercise for the reader to show that the variance derived from the use of the delta technique is given by

$$Var(X_1 X_2) \sim \mu_2^2\,Var(X_1) + \mu_1^2\,Var(X_2) \tag{A3.33}$$

In general a function of several random variables, $F(X_1, X_2, \ldots, X_k)$ will have an approximate variance formula provided by the following:

$$Var\,F(\mathbf{X}) \sim \sum_{i=1}^{k}\left(\frac{\partial F(\ )}{\partial X_i}\right)^2 Var\,X_i + 2\sum_{i=1}^{k}\sum_{\substack{j=1 \\ j>i}}^{k}\frac{\partial F(\ )}{\partial X_i}\frac{\partial F(\ )}{\partial X_j}Cov(X_i, X_j) \tag{A3.34}$$

This expression is equivalent to the matrix equation

$$Var(F(\mathbf{X})) \sim \mathbf{h}V_x\mathbf{h}' \tag{A3.35}$$

where $\mathbf{V}_x$ is the covariance matrix of the vector of random variables $\mathbf{X}' = (X_1, X_2, \ldots, X_k)$ as is given in Equation (A2.24) of Appendix 2 and

$$\mathbf{h} = \left(\frac{\partial F(\ )}{\partial X_1}, \frac{\partial F(\ )}{\partial X_2}, \ldots, \frac{\partial F(\ )}{\partial X_k} \right)$$

is a vector of first derivatives. If there are two functions $F_1(\mathbf{X})$ and $F_2(\mathbf{X})$, then, as before,

$$\text{Var}(F_1(\mathbf{X})) \sim \mathbf{h}_1 \mathbf{V}_x \mathbf{h}_1' \tag{A3.36}$$

and

$$\text{Var}(F_2(\mathbf{X})) \sim \mathbf{h}_2 \mathbf{V}_x \mathbf{h}_2'$$

where $\mathbf{h}_1$ and $\mathbf{h}_2$ are the vectors of first derivatives that correspond to $F_1(\mathbf{X})$ and $F_2(\mathbf{X})$, respectively. Furthermore, if $\mathbf{\mu} = E(\mathbf{X})$, then

$$\text{Cov}[F_1(\mathbf{X}), F_2(\mathbf{X})] \sim E[(F_1(\mathbf{X}) - F_1(\mathbf{\mu}))(F_2(\mathbf{X}) - F_2(\mathbf{\mu}))] \tag{A3.37}$$

$$= E[(F_1(\mathbf{\mu}) + \mathbf{h}_1\mathbf{e} - F_1(\mathbf{\mu}))(F_2(\mathbf{\mu}) + \mathbf{h}_2\mathbf{e} - F_2(\mathbf{\mu}))]$$

$$= E[\mathbf{h}_1\mathbf{e}\mathbf{e}'\mathbf{h}_2']$$

$$= \mathbf{h}_1\mathbf{V}_x\mathbf{h}_2' \tag{A3.38}$$

If one considers u different functions of $\mathbf{X}$, that is

$$\mathbf{F}(\mathbf{X})' = [F_1(\mathbf{X}), F_2(\mathbf{X}), \ldots, F_u(\mathbf{X})] \tag{A3.39}$$

then the covariance matrix of $\mathbf{F}(\mathbf{X})$ is given by

$$\mathbf{V}_F \sim \mathbf{H}\mathbf{V}_\mathbf{X}\mathbf{H}' \tag{A3.40}$$

where, as before, $\mathbf{V}_\mathbf{X}$, is the covariance matrix of $\mathbf{X}$ and $\mathbf{H}$ is a $(u \times k)$ matrix of first partial derivatives of the functions $\mathbf{F}(\mathbf{X})$ evaluated at $\mathbf{\mu}$.

As an example of the use of (A3.40) the covariance matrix of the marginal probabilities derived from a simple 2×2 contingency table will be derived. Let the vector of observed probabilities for this table be represented by

$$\mathbf{p}' = (p_{11}, p_{12}, p_{21}, p_{22}) \tag{A3.41}$$

If the marginal probabilities are represented by the vector

$$\mathbf{m}' = (m_1, m_2, m_3, m_4) = (p_{11} + p_{12}, p_{21} + p_{22}, p_{11} + p_{21}, p_{12} + p_{22})$$

then

$$\mathbf{m} = \mathbf{F}(\mathbf{p}) = \mathbf{A}\mathbf{p} \tag{A3.42}$$

where $\mathbf{A}$ (and $\mathbf{H}$, the matrix of partial derivatives) is given by

$$\mathbf{A} = \begin{bmatrix} 1 & 1 & 0 & 0 \\ 0 & 0 & 1 & 1 \\ 1 & 0 & 1 & 0 \\ 0 & 1 & 0 & 1 \end{bmatrix} \tag{A3.43}$$

Hence

$$\mathbf{V}_m = \mathbf{A}\mathbf{V}_p\mathbf{A}' \tag{A3.44}$$

where $\mathbf{V}_p$ is given in Appendix 1 (Exercise A1.5). Now, on the assumption of the independence of rows and columns, the expected probabilities can be generated by the following vector of functions:

$$\boldsymbol{\pi}' = (m_1 m_3, m_1 m_4, m_2 m_3, m_2 m_4) \tag{A3.45}$$

In this case the matrix of partial derivatives with respect to $\mathbf{m}$ is given by

$$\mathbf{H}^* = \begin{bmatrix} m_3 & 0 & m_1 & 0 \\ m_4 & 0 & 0 & m_1 \\ 0 & m_3 & m_2 & 0 \\ 0 & m_4 & 0 & m_2 \end{bmatrix} \tag{A3.46}$$

and

$$\mathbf{V}_\pi \sim \mathbf{H}^* \mathbf{V}_m \mathbf{H}^* \tag{A3.47}$$

Exercises

A3.1   Provide approximations for the variances of the following functions of X_1, X_2 and X_3 where $E(X_1) = \mu_1$, $E(X_2) = \mu_2$ and $E(X_3) = \mu_3$.

(a) $X_1 X_2$ (b) X_1/X_2 (c) $\log_e(X_1/X_2)$

(d) $X_1 X_2 X_3$ (e) $\dfrac{X_1 X_2}{X_3}$ (f) $X_1 X_2 - X_2 X_3$

(g) $\log_e(X_1) + \log_e(X_2) - \log_e(X_3)$.

A3.2   On the assumption that the four counts in a 3×2 contingency table are distributed according to the multinomial distribution, what is the approximate variance for (a) the natural logarithm of the odds-ratio, and (b) the odds-ratio itself?

A3.3   Koch *et al.* (1977) give the following compounded logarithmic – exponential–linear function for the calculation of a κ coefficient:

$$\kappa = F(\mathbf{p}) = \exp[\mathbf{A}_4(\log_e\{\mathbf{A}_3[\exp(\mathbf{A}_2\{\log_2[\mathbf{A}_1 f]\})]\})] \tag{A3.48}$$

where p is a vector of cell proportions from a square contingency table. They then use the delta method to generate an approximate large-sample variance for κ. On the assumption that you are provided with computer software capable of estimating the variance of a function such as (A3.48), all you would need to provide would be the matrices $\mathbf{A}_1$, $\mathbf{A}_2$, $\mathbf{A}_3$ and $\mathbf{A}_4$. What are they for 2×2 contingency table?

Appendix 4

Computer Software

Most of the methods of statistical analysis that are described in this book have to be performed with the aid of a computer and appropriate statistical software. Many of the required programs are available within general-purpose statistical packages but, unfortunately, there does not seem to be a single package that contains all of them. In addition, there are several techniques such as the use of permutation tests, jackknifing and the use of bootstrap samples which will not usually be catered for as part of a commercial general-purpose package. Here the user will have to write his or her own programs using a language such as FORTRAN or BASIC, for example. There are, however, facilities for the preparation of user-defined subroutines or macros in some of the packages (in SAS, GLIM or GENSTAT, for example – see below).

The following description of software packages is not meant to be complete; it is simply a short review of some of the readily available packages, most of which have been used for preparation of material for this book. No attempt is made here to evaluate the quality of the programs within these packages. Although the packages described below were developed primarily for use on mainframe or minicomputers, most of them are now also available for use on personal computers (PCs).

SPSS^X (Statistical Package for the Social Sciences)

This is a widely used package which contains programs for most of the commonly used statistical methods. The program that might be particularly useful for readers of this text is RELIABILITY. This program is particularly useful for the estimation of several internal consistency measures such as Cronbach's α coefficient. One can also test whether a battery of tests are parallel or τ-equivalent. In the case of a set of binary test responses, bias can be assessed through the use of Cochran's Q statistic. Finally, in some mainframe versions of SPSS^X one can obtain access to LISREL (see below).

Address for enquiries: SPSS Inc.
                                      Suite 3000
                                      444 North Michigan Avenue
                                      Chicago
                                      Illinois 60611
                                      USA

SAS (Statistical Analysis System)

This is a very comprehensive package of statistical programs. SAS is particularly useful for routine data management and manipulation. Users can also write their own programs or subroutines using matrix-handling procedures. In the context of the present text, the two most useful programs (called procedures in SAS) are VARCOMP and CATMOD. VARCOMP is a variance-components estimation program capable of fitting a wide range of random effects or mixed effects models. Estimates can be obtained through the use of both ML and REML criteria as well as through the traditional moments-based (ANOVA) approach. ANOVA estimators of variance components for nested designs can also be obtained through the use of a procedure called NESTED. CATMOD is a procedure for the analysis of categorical measurements. It will enable users to carry out the methods of analysis developed by Grizzle *et al.* (1969) and Koch *et al.* (1977).

Address for enquiries:　　SAS Institute Inc.
　　　　　　　　　　　　　　Box 8000
　　　　　　　　　　　　　　Cary
　　　　　　　　　　　　　　North Carolina 27511-8000
　　　　　　　　　　　　　　USA

BMDP (Biomedical Computer Programs)

This is another general-purpose statistical package with similar capabilities to those of SPSS^X and SAS. Of particular interest here are programs currently called P3V and P8V. These are roughly equivalent to the VARCOMP procedure in SAS.

Address for enquiries:　　BMDP Statistical Software
　　　　　　　　　　　　　　1964 Westwood Boulevard
　　　　　　　　　　　　　　Suite 202
　　　　　　　　　　　　　　Los Angeles
　　　　　　　　　　　　　　California 90025
　　　　　　　　　　　　　　USA

REML (Residual Maximum Likelihood)

REML is a special-purpose package for fitting data to a wide range of mixed effects analysis of variance (and covariance) models using REML as a fitting criterion. A particular useful feature of this package is the way in which it deals with missing values. Using this feature, models can be fitted in which error variances are not constrained to be homogeneous (measuring instruments, for example, can have differing precisions). The reader is referred to D. L. Robinson (1987) for further details.

Address for enquiries:　　Scottish Agricultural Statistical Service
　　　　　　　　　　　　　　University of Edinburgh
　　　　　　　　　　　　　　James Clerk Maxwell Building
　　　　　　　　　　　　　　King's Buildings, Mayfield Road
　　　　　　　　　　　　　　Edinburgh EH9 3JZ
　　　　　　　　　　　　　　UK

LISREL (Analysis of Linear Structural Relationships by Maximum Likelihood)

This is the best-known of the specialised packages for fitting structural relationships models to covariance matrices. It can also be used for both exploratory and confirmatory factor analysis. The reader is referred to Jöreskog and Sörbom (1984) for further details. Users of recent mainframe versions of SPSS^X may be able to call up LISREL from within a SPSS^X run.

Address for enquiries: Scientific Software Inc.
1369 Neitzel Road
Mooresville
IN 46158
USA

GLIM and GENSTAT

These two packages may be of particular use to readers wishing to write their own subroutines for bootstrap or jackknife estimators of standard errors. The bootstrap estimates given in the present book were generated through the use of GLIM macros.

Address for enquiries: Numerical Algorithms Group Ltd
NAG Central Office
Mayfield House
256 Banbury Road
Oxford OX2 7DE
UK

or

Numerical Algorithms Group Inc.
1101 31st Street
Suite 100
Downes Grove
IL 60515-1263
USA

Bootstrap Applications for the Behavioral Sciences

This software provides behavioural scientists with an easy-to-use introduction to Efron's bootstrap approach to the assessment of standard errors and confidence intervals. For further details, readers are referred to

Professor Clifford E. Lunneborg
Mail Stop: NI-25
University of Washington
Seattle
WA 98195
USA

Selected Bibliography

This bibliography includes all references cited in the text together with others that are included for further reading. It is intended that the latter should complement the text for both the statistical methods used in reliability studies and their application in reliability studies by workers in several different disciplines. Readers will find papers on methods that are either not discussed at all in the present book or they are discussed only very briefly. The exploration of patterns of agreement and disagreement through the use of log-linear models, for example, is one area not discussed in this text, but it is an aspect of the analysis of inter-rater reliability studies which is likely to become increasingly important. The selection of references is inevitably a very personal one, but it is hoped that it will be sufficiently broad to satisfy most of the readers of this book.

Allen, M. J. and Yen, W. M. (1979). *Introduction to Measurement Theory*. Brooks/Cole, Monterey.

Allison, P. D. (1978). The reliability of variables measured as a number of events in an interval of time. In K. F. Schuesler (ed.), *Sociological Methodology 1978*, pp. 238–53. Jossey-Bass, San Francisco.

Alwin, D. F. and Jackson, D. J. (1980). Measurement models for response errors in surveys: issues and applications. In K. F. Schuessler (ed.), *Sociological Methodology 1980*, pp. 68–119. Jossey-Bass, San Francisco.

Anderson, D. A. and Aitkin, M. (1985). Variance components models with binary response: interviewer variability. *Journal of the Royal Statistical Society B* **47**, 203–10.

Anderson, R. L. (1987). *Practical Statistics for Analytical Chemists*. Van Nostrand Reinhold, New York.

Andrich, D. (1988). *Rasch Models for Measurement*. Sage Publications, Beverly Hills and London.

Apostol, T. M. (1967). *Calculus Volume 1* (2nd edn). Xerox College Publishing, Lexington, Mass.

Armitage, P. and Berry, G. (1987). *Statistical Methods in Medical Research* (2nd edn). Blackwell Scientific Publications, Oxford.

Armor, D. J. (1974). Theta reliability and factor scaling. In H. C. Costner (ed.), *Sociological Methodology 1973-1974*, pp. 17-50. Jossey-Bass, San Francisco.

Barnett, V. D. (1969). Simultaneous pairwise structural relationships. *Biometrics* **25**, 129-42.

Bartholomew, D. J. (1987). *Latent Variable Models and Factor Analysis*. Charles Griffin, London.

Bartko, J. J. (1966). The intraclass correlation coefficient as a measure of reliability. *Psychological reports* **19**, 3-11.

Bartko, J. J. (1976). On various intraclass correlation reliability coefficients. *Psychological Bulletin* **83**, 762-5.

Bartko, J. J. and Carpenter, W. T. (1976). On the methods and theory of reliability. *The Journal of Nervous and Mental Disease* **163**, 307-17.

Begg, C. B. (1987). Biases in the assessment of diagnostic tests. *Statistics in Medicine* **6**, 411-23.

Bentler, P. M. (1968). Alpha-maximized factor analysis (alphamax). *Psychometrika* **33**, 335-45.

Bentler, P. M. and Chou, C.-P. (1987). Practical issues in structural modelling. *Sociological Methods & Research* **16**, 78-118.

Bishop, Y. M. M., Fienberg, S. E. and Holland, P. W. (1975). *Discrete Multivariate Analysis: Theory and Practice*. MIT Press, Cambridge, Mass.

Boomsma, A. (1985). Nonconvergence, improper solutions, and starting values in LISREL maximum likelihood estimation, *Psychometrika* **50**, 229-42.

Brennan, R. L. (1975). The calculation of reliability from a split-plot factorial design. *Educational and Psychological Measurement* **35**, 779-88.

Brennan, R. L. and Light, R. J. (1974). Measuring agreement when two observers classify people into categories not defined in advance. *British Journal of Mathematical and Statistical Psychology* **27**, 154-63.

Brennan, R. L. and Prediger, D. J. (1981). Coefficient kappa: some uses, misuses and alternatives. *Educational and Psychological Measurement* **41**, 687-99.

British Standards Institute (1987). *British Standard: precision of test methods. Part 1: Guide for the determination of repeatability and reproducibility for a standard test method by interlaboratory tests*. (BS 5487 Part 1).

Brook, R. J. and Stirling, W. D. (1984). Agreement between observers when categories are not specified. *British Journal of Mathematical and Statistical Psychology* **37**, 271-82.

Brown, C. H. (1983). Asymptotic comparison of missing data procedures for estimating factor loadings. *Psychometrika* **48**, 269-91.

Bulmer, M. G. (1979). *Principles of Statistics*. Dover, New York.

Campbell, D. T. and Fiske, D. W. (1959). Convergent and discriminant validation by the multitrait-multimethod matrix. *Psychological Bulletin* **56**, 81-105.

Capobres, D. B., Tosh, F. E., Yates, J. L. and Langeluttig, H. V. (1962). Experience with the tuberculin Tine test in a sanatorium. *Journal of the American Medical Association* **180**, 1130-6.

Carmines, E. G. and Zeller, R. A. (1979). *Reliability and Validity Assessment*. Sage Publications, Beverly Hills and London.

Caulcott, R. and Boddy, R. (1983). *Statistics for Analytical Chemists.* Chapman and Hall, London.

Clare, A. W. and Cairns, V. E. (1978). Design, development and the use of a standardized interview to assess social maladjustment and dysfunction in community studies. *Psychological Measurement* **8**, 589-604.

Cochran, W. G. (1968). Errors of measurement in statistics. *Technometrics* **10**, 637-66.

Cochran, W. G. and Cox, G. M. (1957). *Experimental Designs* (2nd edn). John Wiley and Sons, New York.

Cohen, J. (1960). A coefficient of agreement for nominal scales. *Educational and Psychological Measurement* **20**, 37-46.

Cohen, J. (1968). Weighted kappa: nominal scale agreement with provisions for scales disagreement of partial credit. *Psychological Bulletin* **70**, 213-20.

Colquhoun, D. (1971). *Lectures on Biostatistics.* Oxford University Press, Oxford.

Conger, A. J. (1974). Estimating profile reliability and maximally reliable composites. *Multivariate Behavioral Research* **9**, 85-104.

Conger, A. J. (1980a). Integration and generalization of kappas for multiple raters. *Psychological Bulletin* **88**, 322-8.

Conger, A. J. (1980b). Maximally reliable composites for unidimensional measures. *Educational and Psychological Measurement* **40**, 367-75.

Conger, A. J. and Lipshitz, R. (1973). Measures of reliability for profiles and test batteries. *Psychometrika* **38**, 411-27.

Cox, D. R. (1958). *Planning of Experiments.* John Wiley and Sons, New York.

Cox, D. R. (1970). *Analysis of Binary Data.* Chapman and Hall, London.

Cronbach, L. J. (1951). Coefficient alpha and the internal structure of tests. *Psychometrika* **16**, 297-334.

Cronbach, L. J. (1988). Internal consistency of tests: analyses old and new. *Psychometrika* **53**, 63-70.

Cronbach, L. J., Gleser, G. C., Nanda, H. and Rajaratnam, N. (1972). *The Dependability of Behavioral Measurements.* John Wiley and Sons, New York.

Cronbach, L. J., Rajaratnam, N. and Gleser, G. C. (1963). Theory of generalizability: a liberalization of reliability theory. *British Journal of Statistical Psychology* **16**, 137-63.

Darroch, J. N. and McCloud, P. I. (1986). Category distinguishability and observer agreement. *Australian Journal of Statistics* **28**, 371-88.

Davies, M. and Fleiss, J. L. (1982). Measuring agreement for multinomial data. *Biometrics* **38**, 1047-51.

Dawid, A. P. and Skene, A. M. (1979). Maximum likelihood estimation of observer error-rates using the EM algorithm. *Applied Statistics* **28**, 20-28.

Dayton, C. M. and Macready, G. B. (1983). Latent structure analysis of repeated classifications with dichotomous data. *British Journal of Mathematical and Statistical Psychology* **36**, 189-201.

Dixon, W. J. and Massey, F. J. (1969). *Introduction to Statistical Analysis.* McGraw-Hill, New York.

Donner, A. and Eliasziw, M. (1987). Sample size requirements for reliability studies. *Statistics in Medicine* **6**, 441-8.

Dunn, G. and Everitt, B. S. (1982). *An Introduction to Mathematical Taxonomy*. Cambridge University Press, Cambridge.

Dwyer, J. H. (1983). *Statistical Models for the Social and Behavioral Sciences*. Oxford University Press, New York and Oxford.

Efron, B. (1979). Bootstrap methods: another look at the jackknife. *Annals of Statistics* 7, 1–26.

Efron, B. and Gong, G. (1983). A leisurely look at the bootstrap, the jack-knife, and cross-validation. *The American Statistician* 37, 36–48.

Efron, B. and Tibshirani, R. (1986). Bootstrap methods for standard errors, confidence intervals, and other measures of statistical accuracy. *Statistical Science* 1, 54–77.

Everitt, B. S. (1984). *An Introduction to Latent Variable Models*. Chapman and Hall, London.

Everitt, B. S. and Hand, D. J. (1981). *Finite Mixture Distributions*. Chapman and Hall, London.

Feldstein, M. L. and Davies, H. T. (1984). Poisson models for assessing rater agreement in discrete response studies. *British Journal of Mathematical and Statistical Psychology* 37, 49–61.

Finkbeiner, C. (1979). Estimation for the multiple factor model when data are missing. *Psychometrika* 44, 409–20.

Fleiss, J. L. (1965). Estimating the accuracy of dichotomous judgements. *Psychometrika* 30, 469–79.

Fleiss, J. L. (1970). Estimating the reliability of interview data. *Psychometrika* 35, 143–62.

Fleiss, J. L. (1971). Measuring nominal scale agreement among many raters. *Psychological Bulletin* 76, 378–82.

Fleiss, J. L. (1975). Measuring agreement between judges on the presence or absence of a trait. *Biometrics* 31, 651–9.

Fleiss, J. L. (1981a). Balanced incomplete blocks designs for inter-rater reliability studies. *Applied Psychological Measurement* 5, 105–12.

Fleiss, J. L. (1981b). *Statistical Methods for Rates and Proportions* (2nd edn). John Wiley and Sons, New York.

Fleiss, J. L. (1986). *The Design and Analysis of Clinical Experiments*. John Wiley and Sons, New York.

Fleiss, J. L. and Cohen, J. (1973). The equivalence of weighted kappa and the intraclass correlation coefficient as measures of reliability. *Educational and Psychological Measurement* 33, 613–19.

Fleiss, J. L., Cohen, J. and Everitt, B. S. (1969). Large sample standard errors of kappa and weighted kappa. *Psychological Bulletin* 72, 323–7.

Fleiss, J. L. and Cuzick, J. (1979). The reliability of dichotomous judgements: unequal numbers of judgements per subject. *Applied Psychological Measurement* 3, 537–42.

Fleiss, J. L. and Everitt, B. S. (1971). Comparing marginal totals of square contingency tables. *British Journal of Mathematical and Statistical Psychology* 24, 117–23.

Fleiss, J. L., Spitzer, R. L., Endicott, J. and Cohen, J. (1972). Quantitative agreement in multiple psychiatric diagnosis. *Archives of General Psychiatry* 26, 168–71.

Gillet, R. (1985). Nominal scale response agreement and rater uncertainty. *British Journal of Mathematical and Statistical Psychology* **38**, 58–66.

Gleser, G. C., Cronbach, L. J. and Rajaratnam, N. (1965). Generalizability of scores influenced by multiple sources of variance. *Psychometrika* **30**, 395–418.

Gleser, G. C., Green, B. I. and Winget, C. N. (1978). Quantifying interview data on psychic impairment of disaster survivors. *Journal of Nervous and Mental Disease* **166**, 209–16.

Goldberg, D. P. (1972). *The Detection of Psychiatric Illness by Questionnaire*. Oxford University Press, London.

Goldberg, D. P., Cooper, B., Eastwood, M. R., Kedwood, H. B. and Shepherd, M. (1970). A standardized psychiatric interview for use in community surveys. *British Journal of Preventative and Social Medicine* **24**, 18–23.

Goldstein, H. (1986). Multilevel mixed linear model analysis using iterative generalized least squares. *Biometrika* **73**, 43–56.

Goldstein, H. (1987). *Multilevel Models in Educational and Social Research*. Charles Griffin, London.

Goodman, L. A. (1974). Exploratory latent structure analysis using both identifiable and unidentifiable models. *Biometrika* **61**, 215–31.

Goodman, L. A. and Kruskal, W. H. (1954). Measures of association for cross classifications. *Journal of the American Statistical Association* **49**, 732–64.

Greenberg, R. A. and Jekel, V. F. (1969). Some problems in the determination of the false positive and false negative rates of tuberculin tests. *American Review of Respiratory Disease* **100**, 645–50.

Greene, V. L. and Carmines, E. G. (1979). Assessing the reliability of linear composites. In K. F. Schuessler (ed.), *Sociological Methodology 1980*, pp. 160–75. Jossey-Bass, San Francisco.

Grizzle, V. E., Starmer, C. F. and Koch, G. G. (1969). Analysis of categorical data by linear models. *Biometrics* **25**, 489–504.

Gross, S. T. (1986). The kappa coefficient for multiple observers when the number of subjects is small. *Biometrics* **42**, 883–93.

Grubbs, F. E. (1948). On estimating precision of measuring instruments and product variability. *Journal of the American Statistical Association* **43**, 243–64.

Grubbs, F. E. (1973). Errors of measurement, precision, accuracy and the statistical comparison of measuring instruments. *Technometrics* **15**, 53–66.

Guttman, L. A. (1945). A basis for analysing test retest reliability. *Psychometrika* **10**, 255–82.

Hanamura, R. C. (1975). Measuring imprecisions of measuring instruments. *Technometrics* **17**, 299–302.

Hand, D. J. and Taylor, C. C. (1987). *Multivariate Analysis of Variance and Repeated Measures*. Chapman and Hall, London.

Hartley, H. O., Rao, J. N. K. and LaMotte, S. R. (1978). A simple synthesis-based method of variance components estimation. *Biometrics* **34**, 233–42.

Harville, D. A. (1977). Maximum likelihood approaches to variance component estimation and to related problems. *Journal of the American Statistical Association* **72**, 320–40.

Heiberger, R. M. (1977). Regression with the pairwise-present covariance matrix: a dangerous practice. In *Proceedings of the Statistical Computing Section*, American Statistical Association, Washington.

Heise, D. R. (1969). Separating reliability and stability in test–retest correlation. *American Sociological Review* **34**, 93–101.

Heise, D. R. and Bohrnstedt, G. W. (1970). Validity, invalidity and reliability. In F. Borgatta and G. W. Bohrnstedt (eds), *Sociological Methodology 1970*, pp. 104–29. Jossey-Bass, San Francisco.

Henderson, C. R. (1953). Estimation of variance and covariance components. *Biometrics* **9**, 226–52.

Holmquist, N. D., McMahon, C. A. and Williams, O. D. (1967). Variability in classification of carcinoma in situ of the uterine cervix. *Archives of Pathology* **84**, 334–45.

Hubert, L. and Golledge, R. G. (1983). Rater agreement for complex assessments. *British Journal of Mathematical and Statistical Psychology* **36**, 207–16.

Hubert, L. J. (1977a). Kappa revisited. *Psychological Bulletin* **84**, 289–97.

Hubert, L. J. (1977b). Nominal scale response agreement as a generalized correlation. *British Journal of Mathematical and Statistical Psychology* **30**, 98–103.

Hubert, L. J. (1978). A general formula for the variance of Cohen's weighted kappa. *Psychological Bulletin* **85**, 183–4.

Hui, S. and Walter, S. D. (1980). Estimating the error rates of diagnostic tests. *Biometrics* **36**, 167–71.

Huynh, H. (1986a). On the reliability of an extreme score. *Psychometrika* **51**, 475–8.

Huynh, H. (1986b). Estimation of the KR20 reliability coefficient when the data are incomplete. *British Journal of Mathematical and Statistical Psychology* **39**, 69–78.

Jaech, J. L. (1979). Estimating within laboratory variability from interlaboratory test data. *Journal of Quality Technology* **11**, 185–91.

Jaech, J. L. (1985). *Statistical Analysis of Measurement Errors*. John Wiley and Sons, New York.

Jagodzinski, K. G. and Kühnel, S. M. (1987). Estimation of reliability and stability in single-indicator multiple-wave panel models. *Sociological Methods & Research* **15**, 219–58.

Jagodzinski, K. G., Kühnel, S. M. and Schmidt, P. (1987). Is there a 'Socratic Effect' in nonexperimental panel studies? *Sociological Methods & Research* **15**, 259–302.

Jannarone, R. J., Macera, C. A. and Garrison, C. Z. (1987). Evaluating inter-rater agreement through 'case–control' sampling. *Biometrics* **43**, 433–7.

Jansen, A. A. M. (1980). Comparative calibration and congeneric measurements. *Biometrics* **36**, 729–34.

Jöreskog, K. G. (1971a). Statistical analysis of sets of congeneric tests. *Psychometrika* **36**, 109–35.

Jöreskog, K. G. (1971b). Simultaneous factor analysis in several populations. *Psychometrika* **36**, 409–26.

Jöreskog, K. G. and Goldberger, A. S. (1972). Factor analysis by generalized least squares. *Psychometrika* **37**, 243–59.

Jöreskog, K. G. and Sörbom, D. (1984). *LISREL VI: Analysis of Linear Structural Relationships by the Method of Maximum Likelihood.* Scientific Software Inc., Indiana.

Jørgensen, B. (1985). Estimation of interobserver variation for ordinal rating scales. In R. Gilchrist, B. Francis and J. Whittaker (eds), *Generalized Linear Models*, pp. 93–104. Springer-Verlag, Berlin.

Kendall, M. G. (1938). A new measure of rank correlation. *Biometrika* **30**, 81.

Kendall, M. G. (1943). *The Advanced Theory of Statistics* (5th edn). Charles Griffin, London.

Kjaersgaard-Andersen, P., Christensen, F., Schmidt, S. A., Pedersen, N. W. and Jørgensen, B. (1988). A new method of estimation of interobserver variation and its application to the radiological assessment of osteo-arthrosis in hip joints. *Statistics in Medicine* **7**, 639–47.

Koch, G. G., Landis, J. R., Freeman, J. L., Freeman, D. H. and Lehnen, R. G. (1977). A general methodology for the analysis of experiments with repeated measurement of categorical data. *Biometrics* **33**, 133–58.

Kraemer, H. C. (1976). The small sample non-null properties of Kendall's coefficient of concordance for normal population. *Journal of the American Statistical Association* **71**, 608–13.

Kraemer, H. C. (1979). Ramifications of a population model for κ as a coefficient of reliability. *Psychometrika* **44**, 461–72.

Kraemer, H. C. (1980). Extension of the kappa coefficient. *Biometrics* **36**, 207–16.

Kraemer, H. C. and Korner, A. F. (1976). Statistical alternatives in assessing reliability, consistency and individual differences for quantitative measures: application to behavioral measures of neonates. *Psychological Bulletin* **83**, 914–21.

Krippendorff, K. (1970). Bivariate agreement coefficients for reliability of data. In F. Borgatta and G. W. Bohrnstedt (eds), *Sociological Methodology 1970*, pp. 139–50. Jossey-Bass, San Francisco.

Krippendorff, K. (1978). Reliability of binary attribute data. *Biometrics* **34**, 142–4.

Kristof, W. (1963). Statistical assumptions about error variance. *Psychometrika* **28**, 129–53.

Kristof, W. (1969). Estimation of true score and error variance for tests under various equivalence assumptions. *Psychometrika* **34**, 489–507.

Kristof, W. (1970). On sampling theory of reliability estimation. *Journal of Mathematical Psychology* **7**, 371–7.

Kuder, G. F. and Richardson, M. W. (1937). The theory of the estimation of test reliability. *Psychometrika* **2**, 151–60.

Lachenbruch, P. A. (1988). Multiple reading procedures: the performance of diagnostic tests. *Statistics in Medicine* **7**, 549–57.

LaMotte, L. R. (1973). Quadratic estimation of variance components. *Biometrics* **29**, 311–30.

Landis, J. R. and Koch, G. G. (1975). A review of statistical methods on the analysis of data arising from observer reliability studies. *Statistica Neerlandica* **29**, 101–23 and 151–61.

Landis, J. R. and Koch, G. G. (1977a). The measurement of observer agreement for categorical data. *Biometrics* **33**, 159–74.

Landis, J. R. and Koch, G. G. (1977b). An application of hierarchical kappa-type statistics in the assessment of majority agreement among multiple observers. *Biometrics* **33**, 363–74.

Landis, J. R. and Koch, G. G. (1977c). A one-way components of variance model for categorical data. *Biometrics* **33**, 671–9.

Lawley, D. N. and Maxwell, A. E. (1971). *Factor Analysis as a Statistical Method* (2nd edn). Butterworth, London.

Lazarsfeld, P. F. and Henry, N. W. (1968). *Latent Structure Analysis*. Houghton-Mifflin, New York.

Lee, S.-Y. (1986). Estimation for structural equation models with missing data. *Psychometrika* **51**, 93–9.

Lee, T. D. (1987). Assessment of inter- and intra-laboratory variances: a Bayesian alternative to BS 5497. *The Statistician* **36**, 161–70.

Leone, F. C. and Nelson, L. S. (1966). Empirical studies of variance components I. Empirical studies of balanced nested designs. *Technometrics* **8**, 457–68.

Lewis, G., Pelosi, A. J., Glover, E., Wilkinson, G., Stansfeld, S. A., Williams, P. and Shepherd, M. (1988). The development of a computerized assessment for minor psychiatric disorder. *Psychological Medicine* **18**, 737–45.

Light, R. L. (1971). Measures of response agreement for qualitative data: some generalizations and alternatives. *Psychological Bulletin* **76**, 365–77.

Little, R. J. A. and Rubin, D. B. (1987). *Statistical Analysis with Missing Data*. John Wiley and Sons, New York.

Long, J. S. (1983a). *Confirmatory Factor Analysis: A Preface to LISREL*. Sage Publications, Beverly Hills and London.

Long, J. S. (1983b). *Covariance Structure Models: An Introduction to LISREL*. Sage Publications, Beverly Hills and London.

Lord, F. M. (1980). *Applications of Item Response Theory to Practical Testing Problems*. Laurence Erlbaum Associates, Hillsdale, New Jersey.

Lord, F. M. and Novick, M. R. (1968). *Statistical Theories of Mental Test Scores*. Addison-Wesley, Reading, Mass.

Lundberg, G. A. (1940). The measurement of socioeconomic status. *American Sociological Review* **5**, 29–39.

McCutcheon, A. L. (1988). *Latent Class Analysis*. Sage Publications, Beverly Hills and London.

McHugh, R. B. (1956). Efficient estimation and local identification in latent class analysis. *Psychometrika* **21**, 331–47.

McNemar, Q. (1947). Note on the sampling error of the difference between correlated proportions or percentages. *Psychometrika* **12**, 153–7.

Mandel, J. (1964). *The Statistical Analysis of Experimental Data*. John Wiley and Sons, New York.

Markus, G. B. (1979). *Analysing Panel Data*. Sage Publications, Beverly Hills and London.

Maxwell, A. E. (1970). Comparing the classification of subjects by two independent judges. *British Journal of Psychiatry* **116**, 651–5.

Maxwell, A. E. (1971). Estimating true scores and their reliabilities in the case of composite psychological tests. *British Journal of Mathematical and Statistical Psychology* **24**, 195–204.

Maxwell, A. E. and Pilliner, A. E. G. (1968). Deriving coefficients of reliability and agreement for ratings. *British Journal of Mathematical and Statistical Psychology* **21**, 105–16.

Mezzich, J. E., Kraemer, H. C., Worthington, D. R. C. and Coffman, G. A. (1981). Assessment of agreement among several raters formulating multiple diagnoses. *Journal of Psychiatric Research* **16**, 29–39.

Mosteller, F. and Tukey, J. W. (1977). *Data Analysis and Regression*. Addison-Wesley, New York.

Muthén, B., Kaplan, D. and Hollis, M. (1987). On structural equation modelling with data that are not missing completely at random. *Psychometrika* **52**, 431–62.

Nagelkerke, N. J. D., Fidler, V. and Buwalda, M. (1988). Instrumental variables in the evaluation of diagnostic test procedures when the true disease state is not known. *Statistics in Medicine* **7**, 739–44.

Novick, M. R. and Lewis, C. (1967). Coefficient alpha and the reliability of composite tests. *Psychometrika* **32**, 1–13.

Parzen, E. (1962). *Stochastic Processes*. Holden-Day, San Francisco.

Patterson, H. D. and Thompson, R. (1971). The recovery of inter-block information when block sizes are unequal. *Biometrika* **58**, 545–54.

Patterson, H. D. and Thompson, R. (1975). Maximum likelihood estimation of components of variance. *Proceedings of the 8th International Biometric Conference*, pp. 197–207.

Pearson, K. (1901). Mathematical contributions to the theory of evolution. *Philosophical Transactions of the Royal Society of London (Series A)* **197**, 385–497.

Piaget, J. (1968). *On the Development of Memory and Identity* (Translated by E. Duckworth). Clark University with Barre, Barre, Mass.

Pugh, E. M. and Winslow, G. H. (1966). *The Analysis of Physical Measurements*. Addison-Wesley, Reading, Mass.

Raffalovitch, L. E. and Bohrnstedt, G. W. (1987). Common, specific and error variance components of factor models: estimation with longitudinal data. *Sociological Methods & Research* **15**, 385–405.

Rao, C. R. (1972). Estimation of variance and covariance components in linear models. *Journal of the American Statistical Association* **67**, 112–15.

Rasch, G. (1960). *Probabilistic Models for some Intelligence and Attainment Tests*. Danish Institute for Educational Research, Copenhagen (Expanded edition published by University of Chicago Press, Chicago in 1980).

Rindskopf, D. (1984). Structural equation models: empirical identification. Heywood cases, and related problems. *Sociological Methods & Research* **13**, 109–19.

Rindskopf, D. and Rindskopf, W. (1986). The value of latent class analysis in medical diagnosis. *Statistics in Medicine* **5**, 21–7.

Robinson, D. L. (1987). Estimation and use of variance components. *The Statistician* **36**, 3–14.

Robinson, W. S. (1957). The statistical measurement of agreement. *American Sociological Review* **22**, 17–25.

Rubin, D. B. and Thayer, D. T. (1982). EM algorithms for ML factor analysis. *Psychometrika* **47**, 69–76.

Russell, T. S. and Bradley, R. A. (1958). One-way variances in a two-way classification. *Biometrika* **45**, 111–29.

Scheffé, H. (1959). *The Analysis of Variance*. John Wiley and Sons, New York.

Schouten, H. J. A. (1985). *Statistical Measurement of Interobserver Agreement*. Unpublished doctoral dissertation. Erasmus University, Rotterdam.

Schouten, H. J. A. (1986). Nominal scale agreement among observers. *Psychometrika* **51**, 453–66.

Schwartz, J. E. (1986). A general reliability model for categorical data applied to Guttman scales and current status data. In N. B. Tuma (ed.), *Sociological Methodology 1986*, pp. 79–119. Jossey-Bass, San Francisco.

Scott, W. A. (1955). Reliability of content analysis: the case of nominal scale coding. *Public Opinion Quarterly* **19**, 321–5.

Searle, S. R. (1971). *Linear Models*. John Wiley and Sons, New York.

Searle, S. R. (1987). *Linear Models for Unbalanced Data*. John Wiley and Sons, New York.

Shavelson, R. J. and Webb, N. M. (1981). Generalizability theory: 1973–1980. *British Journal of Mathematical and Statistical Psychology* **34**, 133–66.

Shrout, P. E. and Fleiss, J. L. (1979). Intraclass correlations: uses in assessing rater reliability. *Psychological Bulletin* **86**, 420–8.

Smith, K. W. (1974). On estimating the reliability of composite indexes through factor analysis. *Sociological Methods & Research* **2**, 485–511.

Smith, P. L. (1978). Sampling errors of variance components in small sample multifacet generalizability studies. *Journal of Educational Statistics* **3**, 319–46.

Snedecor, G. W. and Cochran, W. G. (1967). *Statistical Methods* (6th edn). Iowa State University, Ames.

Spearman, C. (1904). General intelligence objectively determined and measured. *American Journal of Psychology* **15**, 201–93.

Spearman, C. (1910). Correlation calculated from faulty data. *British Journal of Psychology* **3**, 271–95.

Stavig, G. R. (1984). Monotonic measures of agreement for ranked data. *British Journal of Mathematical and Statistical Psychology* **37**, 283–7.

Stuart, A. (1955). A test of homogeneity of the marginal distribution in a two-way classification. *Biometrika* **42**, 412–16.

Sullivan, J. L. and Feldman, S. (1979). *Multiple Indicators: An Introduction*. Sage Publications, Beverly Hills and London.

Swallow, W. H. and Monahan, J. F. (1984). Monte Carlo comparison of ANOVA, MIVQUE, REML, and ML estimation of variance components. *Technometrics* **26**, 47–57.

Tanner, M. A. and Young, M. A. (1985). Modelling agreement among raters. *Journal of the American Statistical Association* **80**, 175–80.

Tarnopolsky, A., Hand, D. J., McLean, E. K., Roberts, H. and Wiggins, R. D. (1979). Validity and uses of a screening questionnaire (GHQ) in the community. *British Journal of Psychiatry* **134**, 508–15.

Taylor, J. R. (1982). *An Introduction to Error Analysis*. Oxford University Press, Mill Valley, CA.

Theobald, C. M. and Mallinson, J. R. (1978). Comparative calibration, linear structural relationships and congeneric measurements. *Biometrics* **34**, 39–45.

Thibodeau, L. A. (1981). Evaluating diagnostic tests. *Biometrics* **37**, 801–4.

Thompson, W. A. Jr. (1963). Precision of simultaneous measurement procedures. *Journal of the American Statistical Association* **58**, 474–9.

Thompson, W. O. and Anderson, R. L. (1975). A comparison of designs and estimators for the two-stage nested random model. *Technometrics* **17**, 37–44.

Turner, S. W., Toone, B. K. and Brett-Jones, J. R. (1986). Computerized tomographic scan changes in early schizophrenia – preliminary findings. *Psychological Medicine* **16**, 219–25.

Walter, S. D. (1984). Measuring the reliability of clinical data: the case for using three observers. *Revue Epidémiologie et Santé Publique* **32**, 206–11.

Wechsler, D. (1981). *WAIS-R Manual*. Psychological Corporation, New York.

Werts, C. E., Breland, H. M., Grandy, J. and Rock, D. R. (1980). Using longitudinal data to estimate reliability in the presence of correlated measurement errors. *Educational and Psychological Measurement* **40**, 19–29.

Werts, C. E., Rock, D. R. and Grandy, J. (1979). Confirmatory factor analysis applications: missing data problems and comparison of path models between populations. *Multivariate Behavioral Research* **14**, 199–213.

Werts, C. E., Rock, D. R., Linn, R. L. and Jöreskog, K. G. (1976). Comparison of correlations, variances, covariances, and regression weights with or without measurement error. *Psychological Bulletin* **83**, 1007–13.

Werts, C. E., Rock, D. R., Linn, R. L. and Jöreskog, K. G. (1978). A general method of estimating the reliability of a composite. *Educational and Psychological Measurement* **38**, 933–8.

Wheaton, B., Muthén, B., Alwin, D. F. and Summers, G. F. (1977). Assessing reliability and stability in panel models. In D. R. Heise (ed.), *Sociological Methodology 1977*, pp. 84–136. Jossey-Bass, San Francisco.

Wiley, D. E. and Wiley, J. A. (1970). The estimation of measurement error in panel data. *American Sociological Review* **35**, 112–17.

Williams, G. W. (1976). Comparing the joint agreement of several raters with another rater. *Biometrics* **32**, 619–27.

Woodruff, D. J. and Feldt, L. S. (1986). Tests of equality of several alpha coefficients when their sample estimates are dependent. *Psychometrika* **51**, 393–413.

Yates, F. (1936). Incomplete randomized blocks. *Annals of Eugenics* **7**, 121–40.

Zeller, R. A. (1987). Comment on 'Common, specific and error variance components of factor models: estimation with longitudinal data'. *Sociological Methods & Research* **15**, 406–19.

Zigmond, A. S. and Snaith, R. P. (1983). The Hospital Anxiety and Depression Scale. *Acta Psychiatrica Scandinavica* **67**, 361–70.

Index

Index